APPROXIMATION OF VECTOR VALUED FUNCTIONS

NORTH-HOLLAND
MATHEMATICS STUDIES 25

Notas de Matemática (61)
Editor: Leopoldo Nachbin
*Universidade Federal do Rio de Janeiro
and University of Rochester*

Approximation of Vector Valued Functions

JOÃO B. PROLLA
IMECC, Universidade Estadual de Campinas, Brazil

1977

NORTH-HOLLAND PUBLISHING COMPANY – AMSTERDAM • NEW YORK • OXFORD

North-Holland ISBN: *0 444 85030 9*

PUBLISHERS:
NORTH-HOLLAND PUBLISHING COMPANY
AMSTERDAM • NEW YORK • OXFORD

SOLE DISTRIBUTORS FOR THE U.S.A. AND CANADA:
ELSEVIER NORTH-HOLLAND, INC.
52 VANDERBILT AVENUE, NEW YORK, N.Y. 10017

Library of Congress Cataloging in Publication Data

Prolla, Joao B
Approximation of vector valued functions.

(Notas de matemática ; 61) (North-Holland mathematics studies ; 25)
Bibliography: p.
Includes indexes.
1. Vector valued functions. 2. Approximation theory. I. Title. II. Series.
QA1.N86 no. 61 [QA320] 510'.8s [515'.7]
ISBN 0-444-85030-9 77-22095

PRINTED IN THE NETHERLANDS

PREFACE

This work deals with the many variations of the Stone-Weierstrass Theorem for vector-valued functions and some of its applications. For a more detailed description of its contents see the Introduction and the Table of Contents. The book is largely self-contained. The amount of Functional Analysis required is minimal, except for Chapter 8. But the results of this Chapter are not used elsewhere. The book can be used by graduate students who have taken the usual first-year real and complex analysis courses.

The treatment of the subject has not appeared in book form previously. Even the proof of the Stone-Weierstrass Theorem is new, and due to S. Machado. We also give results in nonarchimedean approximation theory that are new and extend the Dieudonné-Kaplansky Theorem to nonarchimedean vector-valued function spaces.

I thank Professor Silvio Machado, from the Universidade Federal do Rio de Janeiro, for his valuable comments and remarks on the subject. Without his help this would be a different and poorer book. I thank also Professor Leopoldo Nachbin, from the Universidade Federal do Rio de Janeiro and the University of Rochester, whose advice and encouragement was never failing.

Finally, I wish to thank Angelica Marquez and Elda Mortari for typing this monograph.

JOÃO B. PROLLA

Campinas, April 1977

CONTENTS

INTRODUCTION

The typical problem considered in this book is the following. One is given a vector subspace W of a locally convex space L of continuous vector-valued functions, which is a module over an algebra A of continuous scalar-valued functions, and the problem is to describe the closure of W in the space L.

In chapter 1 we start with the case in which $L=C(X;E)$ with the compact-open topology. When the algebra A is self-adjoint, the solution of the above problem is given by the Stone-Weierstrass theorem for modules. A very elegant and direct proof due to S. Machado (see [38]) is presented here. As a corollary one gets the classical Stone-Weierstrass theorem for self-adjoint subalgebras of $C(X;\mathbb{C})$. When the algebra A is not self-adjoint, a solution of the problem is given by Bishop's theorem. The proof that we include here is again due to S. Machado (see [37]). The main idea is to use a "strong" Stone-Weierstrass theorem for the real case plus a transfinite argument. This is done in Machado's paper via Zorn's Lemma. Here we use the transfinite induction process found in the original paper of Bishop (see [8]). We prefer this new method over de Brange's technique, because it can be applied to other situations in weighted approximation theory, namely where measure theoretic tools are either painful to apply or not available at all. In § 9 of this Chapter we treat a special case of vector fibrations, and prove in this context a "strong" Stone-Weierstrass theorem due to Cunningham and Roy (see [15]). This result is used in the next section to characterize extreme functionals. As corollaries, we get the Arens-Kelley theorem for scalar-valued functions, and Singer's theorem (vector-valued case). The results of Buck [12] and Ströbele [63] are also obtained. In an appendix we treat the non locally convex case.

Chapter 2 deals with vector-valued versions of Dieudonné's theorem on the approximation of functions of two variables by means of finite sums of products of functions of one variable (see [18]).

Chapter 3 is devoted to Tietze type extension theorems for vector-valued functions defined on compact subsets of a completely regular Hausdorff space. A typical result says that, if $Y \subset X$ is a compact subset of a completely regular space X, and E is a Fréchet space, then $C_b(X;E)|Y = C(Y;E)$.

The subject matter of chapter 4 is the notion of polynomial algebras. This notion was introduced in Pełczýnski [47], and the name is due to Wulbert (cf. Prenter [49]). In his definition Pełczyńsky used multilinear mappings, whereas Wulbert used polynomials. A third equivalent definition is given in Blatter [4]. We present here Stone-Weierstrass theorem for polynomial algebras. As a corollary we get the infinite dimensional version of the Weierstrass polynomial approximation theorem. Pełczyński attributes this result to S. Mazur (unpublished) in the case of Banach spaces. A much strengthened form of Mazur's result was proved in the joint paper Nachbin, Machado, Prolla [46], namely that the polynomials of finite type from a real locally convex space into another are dense in the space of all continuous function with the compact-open topology. Prenter [48] established Mazur's result for separable

Hilbert spaces. In this chapter we also prove Bishop's theorem for polynomial algebras using the definition given by Pełczyński. It remains an open problem for the more general polynomial algebras. Chapter 4 ends with a study of the approximation of compact linear operators by polynomials of finite type.

In Chapter 5 we are concerned with weighted approximation of vector-valued functions, i.e., with the Bernstein-Nachbin approximation problem. We extend the fundamental work of Nachbin (see for example [43]) from the real or self-adjoint complex case to the general complex case, in the same way that Bishop's theorem generalizes the Stone-Weierstrass theorem. In the joint paper with S. Machado [40], we accomplished this for vector fibrations. Here, however, we restrict ourselves to the particular case of vector-valued functions. As a corollary to our solution of the Bernstein-Nachbin approximation problem we get a strengthened version of Kleinstück's solution of the bounded case (see [35]) of Bernstein-Nachbin problem, as well as of Bishop's theorem for weighted spaces proved by Prolla [51]. The result of Summers [64] for scalar-valued functions is likewise generalized.

In the final two paragraphs of Chapter 5 we study the problem of completeness of Nachbin spaces and the characterization of the dual space of a Nachbin space.

In an appendix to Chapter 5, we present a very simple proof, due to G. Zapata (see [68]), of Mergelyan's theorem characterizing fundamental weights on the real line. This result was then used by Zapata to show that Hadamard's problem on the characterization of quasi-analytic classes of functions is equivalent to Bernstein's problem on the characterization of fundamental weights.

The result of Chapter 5 are applied in Chapter 6 to $C_o(X;E)$, the space of all continuous functions that are E-valued and vanish at infinity on a locally compact space X, equipped with the uniform convergence topology. We also present here Brosowski, Deutsch and Morris theorem (see [10]) on extreme functionals of the unit ball of the dual of $C_o(X;E)$, generalizing it to vector fibrations.

Analogously, in Chapter 7 we apply the results of Chapter 5 to the space $C_b(X;E)$ of all bounded continuous functions, equipped with the strict topology of Buck. We get both Stone-Weierstrass and Bishop's theorem for this topology. We also characterize extreme functionals of polar set of neighborhoods of the origin of $C_b(X;E)$.

The eighth Chapter deals with the ε-product of L. Schwartz and the approximation property for certain spaces of functions, e.g. Aron and Schottenloher [3] result on the equivalence between the approximation property for a complex Banach space E and the same property for the space of holomorphic mappings on E with the compact-open topology. Also, the proof due to K.-D. Bierstedt [5] of the vector-valued version of Mergelyan's theorem on approximation in the complex plane is to be found in this Chapter. It ends with some results of Bierstedt [6] on the "localization" of the approximation property via maximal anti-symmetric sets.

Chapter 9 deals with nonarchimedean approximation Theory. The first results in this area were proved by J. Dieudonné. He proved in [70], for functions with values in the field of p-adic numbers, the analogues of Weierstrass polynomial approximation theorem, and of Stone-Weierstrass Theorem on density of separating subalgebras. To prove these Theorems he first established the analogues of Tietze's Extension Theorem and his own Theorem on appoximation of functions on cartesian products. In 1949, Kaplansky generalized Dieudonné's Stone-Weierstrass Theorem to the case of functions with values in any field with a (rank one) valuation. (See Kaplansky [72]). The case of arbitrary Krull valuations (or of archimedean valuations other than the usual absolute value of $\mathbb{C}$) was established by Chernoff, Rasala and Waterhouse in [69].

We here treat only the case of rank one, i.e. real valued nonarchimedean valuations. We extend the Dieudonné-Kaplansky Theorem to vector valued functions, more precisely to functions with values in a nonarchimedean normed space over some valued field $(F,|\cdot|)$. Our treatment cover the case of A-modules, where A is an algebra of F-valued functions, and in the case $E = F$ extends Kaplansky's result in the sense that we compute the distance of a function from a module. As a corollary one gets the description of the closure of a module and the density result. We also present Murphy's treatment of vector fibrations in a slightly modified version (see [74]). Results on ideals are also given, extending a result of I. Kaplansky on ideals of function algebras (see I. Kaplansky, *Topological Algebra*, Notas de Matemática Nº 16 (Ed. L. Nachbin), Rio de Janeiro.)

CHAPTER 1

THE COMPACT-OPEN TOPOLOGY

§ 1 BASIC DEFINITIONS

Throughout this monograph X denotes a non-void Hausdorff space, and E denotes a non-zero locally convex space over the field $\mathbb{K}$ ($\mathbb{K} = \mathbb{R}$ or $\mathbb{C}$). The topological dual of E is denoted by E', and the set of all continuous seminorms on E is denoted by cs(E).

The vector space over $\mathbb{K}$ of all continuous functions taking X into E is denoted by C(X;E). For every non-void compact subset $K \subset X$ and every continuous seminorm $p \in cs(E)$,

$$f \to ||f||_{K,p} = \sup \{p(f(x));\ x \in K\}$$

defines a seminorm on C(X;E). The topology defined by all such seminorms is called the *compact-open topology*.

When E is a normed space, and $t \to ||t||$ is its norm, we write

$$||f||_K = \sup \{||f(x)||;\ x \in K\}$$

for the corresponding seminorm on C(X;E). In particular, when $E = \mathbb{K}$, we write

$$||f||_K = \sup \{|f(x)|;\ x \in K\}$$

and, if no confusion may arise, $C(X) = C(X;\mathbb{K})$.

The vector subspace of all $f \in C(X;E)$ such that f(X) is a *bounded* subset of E, is denoted by $C_b(X;E)$ and topologized by considering the family of all seminorms

$$f \to ||f||_p = \sup \{p(f(x));\ x \in X\},$$

where $p \in cs(E)$. This topology is referred to as the *topology of uniform convergence* on X, or as the *uniform topology*.

When X is *compact*, the two spaces $C(X;E)$ and $C_b(X;E)$ coincide, and the compact-open and the uniform topology are the same.

When E is a normed space, and $t \to ||t||$ is its norm, we write

$$||f|| = \sup \{||f(x)||;\ x \in X\}$$

for the corresponding norm on $C_b(X;E)$. If $E = \mathbb{K}$, and no confusion may arise, we write $C_b(X) = C_b(X;\mathbb{K})$.

Given a non-empty subset $S \subset C(X;E)$, we define an equivalence relation on X, by setting, for all $x, y \in X$, $x \equiv y$ (mod. S) if, and only if, $f(x) = f(y)$ for all $f \in S$. Since the elements of S are continuous functions, the equivalence classes (mod. S) of X are closed subsets. The set $S \subset C(X;E)$ is said to be *separating on* X if the equivalence classes (mod. S) of X are sets reduced to points. This is equivalent to say that, for any pair $x, y \in X$ of distinct points, there is $f \in S$ such that $f(x) \neq f(y)$. If S is separating on X, we also say that S *separates the points of* X.

If $K \subset X$ is a *closed* non-empty subset, and $S \subset C(X;E)$, then $S|K$ denotes the subset of $C(K;E)$ consisting of all $g \in C(K;E)$ such that there exists $f \in S$ with the property that $g(x) = f(x)$, for all $x \in K$. In particular, if $K \subset X$ is *compact* and $E = \mathbb{K}$, then $C(K) = C_b(X)|K$, by the Tietze Extension Theorem, when X is completely regular.

It follows easily from the above definitions that for any closed subset $K \subset X$, if $x,y \in K$ then $x \equiv y$ (mod. S) if and only if $x \equiv y$ (mod. $S|K$). Moreover, given any equivalence class $Y \subset K$ (mod. $S|K$) there is a *unique* equivalence class $Z \subset X$ (mod. S) such that $Y = Z \cap K$.

Suppose that E is a *Hausdorff space*, and $S \subset C(X;E)$. Let $A = \{\phi \circ f;\ \phi \in E',\ f \in S\}$. Then for every $x,y \in X$, $x \equiv y$ (mod. S) if, and only if, $x \equiv y$ (mod. A). In fact, the "only if" part is true even when E is not Hausdorff.

§ 2 LOCALIZABILITY

Let A be a subalgebra of $C(X;\mathbb{K})$. A vector subspace $W \subset C(X;E)$ will be called a *module over* A, or an A-*module*, if the function $x \to a(x)f(x)$ belongs to W, for every $a \in A$ and $f \in W$.

Notice that, if B denotes the subalgebra of $C(X;\mathbb{K})$ generated by A and the constant functions, then W is an A-module if, and only if, W is a B-module. Moreover, the equivalence relation $x \equiv y$ (mod. A) is the same as $x \equiv y$ (mod. B).

DEFINITION 1.1 *Let* W $C(X;E)$ *be an* A-*module. We say that* W *is localizable under* A *in* $C(X;E)$ *if the compact-open closure of* W *in* $C(X;E)$ *is the set of all* $f \in C(X;E)$ *such that* $f|Y$ *belongs to the compact-open closure of* $W|Y$ *in* $C(Y;E)$ *for each equivalence class* $Y \subset X$ *(mod.* A*).*

This is equivalent to say that the compact-open closure of W in $C(X;E)$ is the set of all $f \in C(X;E)$ such that, given $Y \subset X$ an equivalence class (mod. A), $K \subset Y$ a compact subset, $\varepsilon > 0$; and $p \in cs(E)$, there is $g \in W$ such that $p(f(x)-g(x)) < \varepsilon$, for all $x \in K$. We let $L_A(W)$ be the set of all such functions. Notice that $L_A(W)$ is always a closed subset of $C(X;E)$, containing W. It follows that W is localizable under A in $C(X;E)$ if, and only if, $L_A(W)$ is contained in the compact-open closure $\overline{W}$ of W in $C(X;E)$.

Notice too that $L_A(W) = L_B(W)$, if B denotes the subalgebra of $C(X;\mathbb{K})$ generated by A and the constant functions. Thus W is localizable under A in $C(X;E)$ if, and only if, W is localizable under B in $C(X;E)$.

When $E = \mathbb{K}$, every subalgebra $A \subset C(X;\mathbb{K})$ is a module over itself. The definition of localizability is motivated by the classical Stone-Weierstrass Theorem. Indeed, we have the following result which connects the notion of localizability with the usual statement of the Stone-Weierstrass Theorem. (See Theorem 1, § 17, Nachbin [43]).

PROPOSITION 1.2 *Let* $A \subset C(X;\mathbb{K})$ *be a* $\mathbb{K}$-*subalgebra, and let* $f \in C(X;\mathbb{K})$. *Then* $f \in L_A(A)$ *if, and only if, the following two conditions are satisfied:*

(1) *for every* $x \in X$ *such that* $f(x) \neq 0$, *there exists* $g \in A$ *such that* $g(x) \neq 0$;

(2) *for every* $x,y \in X$ *such that* $f(x) \neq f(y)$, *there exists* $g \in A$ *such that* $g(x) \neq g(y)$.

PROOF (a) Suppose $f \in L_A(A)$. Let $x \in X$ be such that $f(x) \neq 0$. Assume that $g(x) = 0$ for all $g \in A$. Let $Y \subset X$ be the equivalence class (mod. A) that contains x, and let $K = \{x\}$. By hypothesis, there is $g \in A$ such that $|f(x) - g(x)| < \varepsilon = |f(x)|$. Since $g(x) = 0$, this is a contradiction. Therefore (1) is satisfied. The proof that (2) is satisfied is analogous, so we omit the details.

Suppose now conditions (1) and (2) are satisfied. Let $Y \subset X$ be an equivalence class (mod. A). By (2), f is constant on Y. Let $u \in \mathbb{K}$ be its constant value. If $u = 0$, then $g = 0 \in A$ coincides with f on Y. Assume now that $u \neq 0$. By (1), there is $g \in A$ such that $g(x) \neq 0$, where $x \in Y$ is an arbitrary point fixed in Y. Then $g(y) = g(x)$ for all $y \in Y$. Therefore $h = (u/g(x))g$ belongs to A and $h(y) = u = f(y)$ for all $y \in Y$. Hence $f \in L_A(A)$.

§ 3 PRELIMINARY LEMMAS

In this section we shall obtain two lemmas that will be useful in the "approximate partition of unity" needed in the proof of the main theorem of this chapter. The second of those lemmas is due to Jewett [32], who employed it in his proof of a variation of the Stone-Weierstrass theorem. It is a corollary of the classical Weierstrass polynomial approximation theorem, but we prefer to present Jewett's direct proof, to make our version of the Stone-Weierstrass theorem independent of Weierstrass theorem.

LEMMA 1.3 *Let* $A \subset C_b(X;\mathbb{R})$ *be a subalgebra containing the constants, and let* $Y \subset X$ *be an equivalence class (mod. A). For every* $\varepsilon > 0$, *and every open subset* $U \subset X$, *containing* Y, *such that the complement of* U *is compact, we can find* $g \in A$ *such that* $0 \leq g \leq 1$, $g(y) = 1$ *for all* $y \in Y$, *and* $g(t) < \varepsilon$ *for* $t \notin U$.

PROOF Choose $x \in Y$. For each $f \in A$, let $X_f = \{t \in X; f(t) = f(x)\}$. Then

$$Y = \bigcap \{X_f;\ f \in A,\ f \text{ is not constant}\}$$

The compact set $X \setminus U$ is covered by the family of open sets $\{X \setminus X_f;\ f \in A,\ f \text{ is not constant}\}$. By compactness, we can find a finite number of functions $f_1, \ldots, f_n \in A$, none of then being constant, such that

$$(X \setminus U) \subset (X \setminus X_{f_1}) \cup \ldots \cup (X \setminus X_{f_n}).$$

Consequently, $Y \subset (X_{f_1} \cap \ldots \cap X_{f_n}) \subset U$.

For each $i = 1, \ldots, n$, define $g_i \in A$ by

$$g_i = (f_i - f_i(x))^2 / 2\, \|f_i - f_i(x)\|^2.$$

Then $0 \leq g_i < 1$, and $h_i = 1 - g_i$ belongs to the algebra A, $0 < h_i \leq 1$, and $h_i(y) = 1$ for $y \in X_{f_i}$, and $h_i(t) < 1$ for t in the complement of X_{f_i}. Define

$$g_0 = (h_1 + \ldots + h_n)/n.$$

Then $g_0 \in A$, $0 < g_0 \leq 1$, $g_0(y) = 1$ for $y \in X_{f_1} \cap \ldots \cap X_{f_n}$, a fortiori for $y \in Y$. If $t \notin U$, then $t \notin X_{f_i}$ for some i. Hence $g_0(t) < 1$. Since the complement of U is compact,

$$m = \sup \{g_0(t);\ t \notin U\} < 1.$$

For $k \in \mathbb{N}$ sufficiently large, $m^k < \varepsilon$. Then $g = g_0^k$ belongs to A; $0 < g \leq 1$; $g(y) = 1$ for all $y \in Y$; and $g(t) < \varepsilon$ for all $t \notin U$, as desired.

LEMMA 1.4 *Let* $0 < \varepsilon < 1 - \varepsilon$. *Given* $\delta > 0$, *there exists a polynomial* $q: \mathbb{R} \to \mathbb{R}$ *such that*

(a) $0 \leq q(t) \leq 1$, *for all* $0 \leq t \leq 1$;

(b) $0 \leq q(t) < \delta$, *for all* $0 \leq t \leq \varepsilon$;

(c) $1 - \delta < q(t) \leq 1$, *for all* $1 - \varepsilon \leq t \leq 1$.

PROOF (Jewett [32]) The polynomial $q\colon \mathbb{R} \to \mathbb{R}$ will be of the form $q(t) = 1-(1-t^m)^n$. Choose an integer r such that $(3/4)^r < \delta$. Then choose integers m and s such that

$$\frac{3}{4}\cdot\frac{1}{(1-\varepsilon)^m} < s < \frac{1}{(1-\varepsilon)^m} < \frac{\delta}{r}\cdot\frac{1}{\varepsilon^m} .$$

Let $n = rs$ and note that $n\varepsilon^m < \delta$, and $(3/4) < s(1-\varepsilon)^m < 1$. Hence

$$\text{(i)}\quad (1-\varepsilon^m)^n \geq 1 - n\varepsilon^m > 1 - \delta;$$

$$\text{(ii)}\quad (1-(1-\varepsilon)^m)^n = [(1-(1-\varepsilon)^m)^s]^r < \left[1 - s(1-\varepsilon)^m + \tfrac{1}{2}(s(1-\varepsilon)^m)^2\right]^r < \left(\frac{3}{4}\right)^r < \delta.$$

Let $p(t) = (1-t^m)^n$. From (i), it follows that $p(t) > 1-\delta$, for all $0 \leq t \leq \varepsilon$. Hence $q(t) = 1 - p(t) < \delta$, for all $0 \leq t \leq \varepsilon$. From (ii), it follows that $p(t) < \delta$ for all $1-\varepsilon \leq t \leq 1$. Hence $q(t) > 1-\delta$, for all $1-\varepsilon \leq t \leq 1$, and the present lemma is true.

We now establish the main approximation theorem for vector-valued continuous functions, namely a Stone-Weierstrass theorem for modules. The proof is due to S. Machado [38]. When $E = \mathbb{K}$, it was proved by L. Nachbin (see Theorem 1, § 19,[40]), using the classical Stone-Weierstrass theorem for algebras, which is in turn a corollary of the theorem for modules established by Nachbin. To see this, it suffices to notice that, by proposition 1.2, the classical Stone-Weierstrass theorem states that every subalgebra $A \subset C(X;\mathbb{R})$ is localizable under itself in $C(X;\mathbb{R})$. However, Machado's proof, relying only on Lemmas 1.3 and 1.4, provides in particular a very elegant and direct proof of the Stone-Weierstrass theorem. We finally remark that the fact that we are dealing with vector-valued functions, causes no additional difficulty. Indeed the proof for $E = \mathbb{K}$ would just substitute estimates with absolute value for estimates with seminorms on E.

§ 4 STONE-WEIERSTRASS THEOREM FOR MODULES

THEOREM 1.5 $A \subset C(X;\mathbb{R})$ *be a subalgebra. Every* A-*module* $W \subset C(X;E)$ *is localizable under* A *in* $C(X;E)$.

PROOF Since $L_A(W) = L_B(W)$, and W is an A-module if, and only if, it is a B-module, where B denotes the subalgebra of $C(X;\mathbb{R})$ generated by the A and the constants, we may assume without loss of generality that A contains the constants. Moreover, the case of a general X follows easily from the case of X compact, since for any compact set $K \subset X$, if $Y \subset K$ is an equivalence class (mod. $A|_K$), then there exists a unique equivalence class $Z \subset X$ (mod. A) such that $Y = Z \cap K$, and Y is compact subset of Z. Hence, we may also assume without loss of generality that X is compact.

Let then $f \in L_A(W)$. We claim that $f \in \overline{W}$. Let $0<\varepsilon<1$ and $p \in cs(E)$ be given. For every equivalence class $Y \subset X$ (mod. A), let $w_Y \in W$ be such that $p(f(x) - w_Y(x)) < \varepsilon/3$ for all $x \in Y$. Then

$$U_Y = \{t \in X;\ p(f(t)- w_Y(t))< \varepsilon/3\}$$

is an open subset of X, containing Y. By Lemma 1.3 there is $g_Y \in A$, such that $0 \le g_Y \le 1$, $g_Y(y) = 1$ for all $y \in Y$, and $g_Y(t) < \varepsilon/3$ for all $t \notin U_Y$. Let

$$V_Y = \{t \in X;\ g_Y(t) > 1 - (\varepsilon/3)\}.$$

Then V_Y is open and contains Y. Moreover $V_Y \subset U_Y$. Indeed, if $t \notin U_Y$, then $g_Y(t) < \varepsilon/3$. If $t \in V_Y$ were true, then $(\varepsilon/3) > 1 - (\varepsilon/3)$, and this contradicts $\varepsilon < 1$. This proves our claim. By compactness of X, there exist $Y_1,\dots,Y_n$ equivalence classes (mod. A) such that $X = V_1 \cup \dots \cup V_n$, where for each $i=1,\dots,n$, V_i denotes V_Y, for $Y = Y_i$. Let $0 < \delta < (3n(M+1))^{-1}\varepsilon$, where M is the constant $\max\{\|f\|_p, \|f - w_1\|_p,\dots,\|f - w_n\|_p\}$. To simplify notation, we have written $w_i = w_Y$, for $Y = Y_i$.

By Lemma 1.4, there is a polynomial $q\colon \mathbb{R} \to \mathbb{R}$ such that

(a) $0 \le q(t) \le 1$, for all $0 \le t \le 1$;

(b) $0 \leq g(t) < \delta$, for all $0 \leq t \leq \varepsilon/3$;

(c) $1 - \delta < g(t) \leq 1$, for all $1-(\varepsilon/3) \leq t \leq 1$.

Let $g_i = g \circ g_Y$, for $Y = Y_i$ $(i=1,2,\ldots,n)$. Then

$$g_i \in A,\ 0 \leq g_i \leq 1 \text{ and}$$

(1) $0 \leq g_i(t) < \delta$, if $t \notin U_i$

(2) $1 - \delta < g_i(x) \leq 1$, if $x \in V_i$

for $i=1,2,\ldots,n$, and $U_i = U_Y$, for $Y = Y_i$. Following Rudin [55], item 2.13, let us define

$$h_1 = g_1,\ h_2 = (1 - g_1)g_2, \ldots,$$

$$h_n = (1-g_1)(1-g_2)\ldots(1-g_{n-1})g_n. \text{ Then}$$

(3) $h_1 + h_2 + \ldots + h_n = 1-(1-g_1)(1-g_2)\ldots(1-g_n)$.

In view of $0 \leq g_i \leq 1$ and $0 \leq 1-g_i \leq 1$, we see that $0 \leq h_i \leq 1$, for all $i=1,2,\ldots,n$. Given $x \in X$, there is some index i such that $x \in V_i$. By (2), $g_i(x) > 1 - \delta$. We now use (3) and obtain

$$(4)\quad 1 \geq \sum_{i=1}^{n} h_i(x) = 1-(1-g_i(x)) \prod_{j \neq i} (1-g_i(x)) > 1 - \delta.$$

On the other hand, $h_i(t) \leq g_i(t)$ and formula (1) lead to

(5) $0 \leq h_i(t) < \delta$, if $t \notin U_i$

for all $i=1,2,\ldots,n$.

Let $w = \sum_{i=1}^{n} h_i w_i \in W$. For each $x \in X$, we have

$$p(f(x)-w(x)) \leq p(f(x) - \sum_{i=1}^{n} h_i(x) f(x)) +$$

$$+ p(\sum_{i=1}^{n} h_i(x)(f(x) - w_i(x))).$$

For the first term in the right-hand side, we obtain, by (4), the estimate:

$$p(f(x)-\sum_{i=1}^{n} h_i(x)f(x)) \leq |1-\sum_{i=1}^{n} h_i(x)| \cdot ||f||_p$$

$$< \delta||f||_p < \varepsilon/3.$$

To evaluate the second term, let

$I_x = \{1 \leq i \leq n;\ x \in U_i\}$ and $J_x = \{1 \leq i \leq n; x \notin U_i\}$.

Then, for $i \in I_x$, $p(h_i(x)(f(x)-w_i(x))) < h_i(x)(\varepsilon/3)$;

and for $i \in J_x$, $p(h_i(x)(f(x)-w_i(x))) < \delta \cdot ||f-w_i||_p < M\delta$, by using (5). Hence

$$p(f(x)-w(x)) < \frac{\varepsilon}{3} + \sum_{i=1}^{n} h_i(x)p(f(x)-w_i(x))$$

$$= \frac{\varepsilon}{3} + \Sigma' + \Sigma'' < \frac{\varepsilon}{3} + \frac{\varepsilon}{3} + n\,M\,\delta < \varepsilon,$$

where Σ' is the sum for $i \in I_x$ and Σ'' is the sum for $i \in J_x$.

Therefore $f \in \overline{W}$, as claimed, and that concludes the proof.

COROLLARY 1.6 *(Stone-Weierstrass theorem). Let* $A \subset C(X;\mathbb{R})$ *be a subalgebra and let* $f \in C(X;\mathbb{R})$. *Then* f *belongs to the closure of* A *in* $C(X;\mathbb{R})$ *if, and only if, the following two conditions are satisfied:*

(1) *for every* $x \in X$ *such that* $f(x) \neq 0$, *there is* $g \in A$ *such that* $g(x) \neq 0$;

(2) *for every* $x, y \in X$ *such that* $f(x) \neq f(y)$, *there is* $g \in A$ *such that* $g(x) \neq g(y)$.

PROOF Since every algebra is a module over itself, $\overline{A} = L_A(A)$, by Theorem 1.5. On the other hand, by Proposition 1.2, with $\mathbb{K} = \mathbb{R}$, we have that $f \in L_A(A)$ if, and only if, conditions (1) and (2) are satisfied.

§ 5 THE COMPLEX SELF-ADJOINT CASE

The following lemma is the key to the reduction of the complex self-adjoint case to the real case.

LEMMA 1.7 *Let* $A \subset C(X;\mathbb{C})$ *be a self-adjoint subalgebra, and let* E *be a locally convex vector space over* $\mathbb{C}$. *Let* $B \subset C(X;\mathbb{R})$ *be the set* $\{\text{Re } f;\ f \in A\}$. *Then*

(1) $B \subset A$;

(2) B *is a subalgebra of* $C(X;\mathbb{R})$.

For every vector subspace $W \subset C(X;E)$ *we have:*

(3) W *is an* A*-module if, and only if,* W *is a* B*-module;*

(4) $L_A(W) = L_B(W)$.

PROOF For every $f \in A$, $\text{Re } f = (f+\bar{f})/2$. Hence $\bar{f} \in A$ implies $\text{Re } f \in A$, which proves (1).

Clearly, B is a vector subspace of $C(X;\mathbb{R})$. Now, if $f, g \in A$, then by (1), $\text{Re } f$ and $\text{Re } g$ belong to A. Hence $(\text{Re } f)(\text{Re } g) \in A$. Since $(\text{Re } f)(\text{Re } g)$ is real-valued, it follows that $(\text{Re } f)(\text{Re } g) \in B$. Thus (2) holds.

Let now W be a vector subspace of $C(X;E)$. Since $B \subset A$, W is an A-module implies that W is a B-module. To prove the converse, it is sufficient to prove that $A \subset B + i\,B$. Let $f \in A$, $f = u + i\,v$. By definition, $u \in B$. On the other hand $v = \text{Re } g$, where $g = -i \cdot f$. Hence $v \in B$, and the proof of (3) is complete.

Finally, (4) follows from the fact that $x \equiv y$ (mod. A) if and only if $x \equiv y$ (mod. B). To prove this last fact, notice that $x \not\equiv y$ (mod. A) implies the existence of $f \in A$ such that $f(x) \neq f(y)$. If $\text{Re } f(x) \neq \text{Re } f(y)$, then $x \not\equiv y$ (mod. B). If on the other hand, $\text{Im } f(x) \neq \text{Im } f(y)$, then $g = -i \cdot f$ is such that $g \in A$ and $\text{Re } f(x) \neq \text{Re } g(y)$. Hence $x \not\equiv y$ (mod. B) in this case too. The converse, $x \equiv y$ (mod. A) $\Rightarrow$ $x \equiv y$ (mod. B), follows from $B \subset A$. This completes the proof of the present lemma.

THEOREM 1.8 *Let* $A \subset C(X;\mathbb{C})$ *be a self-adjoint subalgebra, and let* E *be a locally convex space over* $\mathbb{C}$. *Then every* A*-submodule* $W \subset C(X;E)$ *is localizable under* A *in* $C(X;E)$.

PROOF Let $B = \{\mathrm{Re}\, f;\ f \in A\}$. By Lemma 1.7, W is a B-module. By theorem 1.5, W is localizable under B in $C(X;E)$. Hence $L_B(W) = \overline{W}$. However, by part (4) of Lemma 1.7, $L_B(W) = L_A(W)$. Therefore, $L_A(W) = \overline{W}$, as desired.

COROLLARY 1.9 *Let $A \subset C(X;\mathbb{C})$ be a self-adjoint subalgebra, and let $f \in C(X;\mathbb{C})$. The following statements are equivalent.*

(1) *f belongs to the closure of A;*

(2) *given $x, y \in X$ and $\varepsilon > 0$, there is $g \in A$ such that $|g(x) - f(x)| < \varepsilon$, $|g(y) - f(y)| < \varepsilon$;*

(3) (a) *for every $x, y \in X$ such that $f(x) \neq f(y)$, there is $g \in A$ such that $g(x) \neq g(y)$; and*

(b) *for every $x \in X$ such that $f(x) \neq 0$, there is $g \in A$ such that $g(x) \neq 0$;*

(4) *f belongs to $L_A(A)$.*

PROOF (1) ⇒ (2). Obvious.

(2) ⇒ (3). Let $x, y \in X$ be such that $f(x) \neq f(y)$. Define $\varepsilon = |f(x) - f(y)|$. Then $\varepsilon > 0$, and by (2), there is $g \in A$ such that $|g(x) - f(x)| < \varepsilon/2$ and $|g(y) - f(y)| < \varepsilon/2$.

If $g(x) = g(y)$, then

$$\varepsilon = |f(x)-f(y)| = |f(x)-g(x)+g(y)-f(y)| \leq$$

$$\leq |f(x)-g(x)| + |g(y)-f(y)| < \varepsilon/2 + \varepsilon/2 = \varepsilon,$$

a contradiction. This proves part (a) of (3). A similar argument proves part (b).

(3) ⇒ (4), by Proposition 1.2.

(4) ⇒ (1), by Theorem 1.8, since every algebra is a module over itself.

COROLLARY 1.10 *Let $A \subset C(X;\mathbb{C})$ be a closed self-adjoint subalgebra, and let $f \in C(X;\mathbb{C})$. Then $f \in A$ if, and only if, given $x, y \in X$ there is $g \in A$ such that $g(x) = f(x)$ and $g(y) = f(y)$.*

REMARK. For further results see Arens [1].

We will finish this paragraph by presenting two applications of Theorem 1.8.

EXAMPLE I. Let X be the closed interval $[0,1] \subset \mathbb{R}$. Let

$$L = \{ f \in C([0,1] ; \mathbb{C}); f(0) = 0 \} .$$

Let W_1 be the vector subspace of $C([0,1] ; \mathbb{C})$ generated by the functions $t \mapsto t^{2n+1}$ for $n = 0,1,2,3,\dots$. Clearly, W_1 is not a subalgebra of $C([0,1] ;\mathbb{C})$. However, W_1 is an A-module, where A is the vector space generated by the functions $t \mapsto t^{2n}$ for $n = 0,1,2,3,\dots$. In fact, A is a separating subalgebra of $C([0,1];\mathbb{C})$, which is self-adjoint and contains the constants. By Theorem 1.8, $f \in C([0,1] ;\mathbb{C})$ belongs to $\overline{W_1}$ if, and only if, $f(x) \in \overline{W_1(x)}$ for each $x \in [0,1]$. Now $W_1(x) = 0$ if $x = 0$, and $W_1(x) = \mathbb{C}$ if $0 < x \leq 1$. Hence $\overline{W_1} = L$ in $C([0,1] ;\mathbb{C})$.

Let now W_2 be the vector subspace of $C([0,1] ;\mathbb{C})$ generated by the functions $t \mapsto t^{2n}$, for $n = 1,2,3,\dots$. W_2 is an algebra, but we prefer to look at the fact that W_2 is also an A-module. Again $W_2(x) = 0$ if $x = 0$, and $W_2(x) = \mathbb{C}$ if $0 < x \leq 1$. By Theorem 1.8, we have

$$\overline{W_2} = L \quad \text{in} \quad C([0,1]; \mathbb{C})$$

EXAMPLE II. Let E be a complex Banach space with norm $\|\cdot\|$, and let $X = \mathbb{R}^n$. Let $\mathcal{E}(\mathbb{R}^n ;\mathbb{C})$ and $\mathcal{E}(\mathbb{R}^n ;E)$ be the vector spaces of indefinitely differentiable functions on $\mathbb{R}^n$ with values in $\mathbb{C}$ and E, respectively. Then $\mathcal{E}(\mathbb{R}^n; \mathbb{C}) \subset C(\mathbb{R}^n ;\mathbb{C})$ is a separating subalgebra which is self-adjoint and contains the constants. The same assertions are true with respect to the algebra $\mathcal{D}(\mathbb{R}^n ;\mathbb{C})$ consisting of all $f \in \mathcal{E}(\mathbb{R}^n; \mathbb{C})$ with compact support. Both the spaces $\mathcal{E}(\mathbb{R}^n; E)$ and $\mathcal{D}(\mathbb{R}^n; E)$ are vector subspaces of $C(\mathbb{R}^n; E)$ which are A-modules for $A = \mathcal{E}(\mathbb{R}^n; \mathbb{C})$ or $A = \mathcal{D}(\mathbb{R}^n; \mathbb{C})$. Here $\mathcal{D}(\mathbb{R}^n; E)$ is the vector space of all $f \in \mathcal{E}(\mathbb{R}^n ; E)$ which have compact support.

Now for each $x \in \mathbb{R}^n$, there is some $\phi_x \in \mathcal{D}(\mathbb{R}^n ; \mathbb{C})$ such that $\phi_x(x) = 1$. Hence $W(x) = E$, for $W = \mathcal{E}(\mathbb{R}^n ;E)$ or for $W = \mathcal{D}(\mathbb{R}^n ;E)$. By Theorem 1.8, both $\mathcal{E}(\mathbb{R}^n;E)$ and $\mathcal{D}(\mathbb{R}^n;E)$ are dense in the compact-open topology of $C(\mathbb{R}^n; E)$.

§ 6 SUBMODULES OF C(X;E)

Let us begin with the following corollary to theorem 1.8.

THEOREM 1.11 *Let* $A \subset C(X)$ *be a separating self-adjoint subalgebra and let* $W \subset C(X;E)$ *be a vector subspace which is an* A*-module. A function* $f \in C(X;E)$ *belongs to the closure of* W, *if and only if,* f(x) *belongs to the closure of* $W(x) = \{g(x); g \in W\}$ *in* E, *for each* $x \in X$.

PROOF Since A is separating, each equivalence class $Y \subset X$ (mod. A) is a set reduced to a point, $Y = \{x\}$, and $W|Y = W(x) = \{g(x);\ g \in W\} \subset E$.

Using the above theorem we can prove a result on ideals in function algebras, due to I. Kaplansky. If E is a locally convex space endowed with a jointly continuous multiplication, then C(X;E) becomes an algebra (over the same field of E) when we define operations pointwise. Now the problem arises of characterizing the closed right (resp. left) ideals $I \subset C(X;E)$. Suppose that for every $x \in X$ a closed right (resp. left) ideal $I_x \subset E$ is given, and let $I = \{f \in C(X;E);\ f(x) \in I_x \text{ for all } x \in X\}$. Clearly, I is a closed right (resp. left) ideal in C(X;E). We shall prove that *any* closed right (resp. left) ideal in C(X;E) has the above form, if E has a unit.

THEOREM 1.12 (Kaplansky) *Let* E *be a locally convex space endowed with a jointly continuous multiplication. Assume that* E *has a unit. Let* $I \subset C(X;E)$ *be a closed right (resp. left) ideal. For each* $x \in X$, *let* I_x *be the closure of* I(x) *in* E. *Then* I_x *is a closed right (resp. left) ideal in* E, *and* $I = \{f \in X(X;E); f(x) \in I_x$ *for all* $x \in X\}$.

PROOF For every $x \in X$, I(x) is clearly a right (resp. left) ideal in E. Since the multiplication in E is jointly continuous, the closure of any right (resp. left) ideal in E is a right (resp. left) ideal. Hence I_x is a right (resp. left) ideal in E. We claim that I is a C(X)-module. Indeed, if $f \in I$ and $g \in C(X)$, let $h \in C(X;E)$ be the function $x \mapsto g(x)e$, where e is the unit of E. If I is a right ideal, then

$$g(x)f(x) = g(x)[f(x)e] = f(x)[g(x)e] = f(x)\ h(x)$$

for all $x \in X$, and therefore $g\ f = f\ h \in I$. The case of a left ideal is treated similarly, by writing

$$g(x)f(x) = g(x)[e\ f(x)] = [g(x)e]f(x) = h(x)\ f(x).$$

It remains to apply Theorem 1.11 to the separating (self-adjoint in the complex case) algebra $A = C(X)$, and the closed A-module I.

COROLLARY 1.13 *Under the hypothesis of Theorem 1.12 assume that the algebra* E *is simple. Then any closed two-sided ideal consists of all functions vanishing on a closed subset of* X.

PROOF We first recall that E is said to be *simple* if it has no two-sided ideals other than 0 and E. Let $N \subset X$ be a closed subset of X. Clearly, $Z(N) = \{f \in C(X;E);\ f(x) = 0$ for all $x \in N\}$ is a closed two-sided ideal of $C(X;E)$.

Conversely, if I is a closed two-sided ideal in $C(X;E)$, let us define $N = \{x \in X;\ f(x) = 0$ for all $f \in I\}$. Clearly, N is closed in X and $I \subset Z(N)$. To prove the converse, i.e. to prove $Z(N) \subset I$, let $f \in Z(N)$, and assume by contradiction that $f \notin I$. By Theorem 1.12, there is an $x \in X$ such that $f(x) \notin I_x$. Since I_x is a two-sided ideal, $I_x = \{0\}$; the case $I_x = E$ being impossible because $f(x) \in E$. Hence $f(x) \neq 0$. Since $f \in Z(N)$, $x \notin N$. However, $I_x = \{0\}$ implies $I(x) = 0$, and so $x \in N$, a contradiction.

COROLLARY 1.14 *Let* A *and* W *be as in Theorem 1.11. Then* W *is dense in* $C(X;E)$ *if, and only if,* $W(x)$ *is dense in* E*, for each* $x \in X$.

We shall denote by $C(X) \otimes E$, the vector subspace of $C(X;E)$ consisting of all finite sums of functions of the form $x \to f(x)v$, where $f \in C(X)$ and $v \in E$. Clearly, $C(X) \otimes E$ is a $C(X)$-module.

THEOREM 1.15 *Let* X *be a Hausdorff space such that* $C(X)$ *separates the points of* X*. Then* $C(X) \otimes E$ *is dense in* $C(X;E)$.

PROOF Let $W = C(X) \otimes E$. By hypothesis, W is module over a self-adjoint separating subalgebra, namely $C(X)$. Moreover, $W(x) = E$, for each $x \in X$. The result now follows from Corollary 1.14 above.

Theorem 1.15 above can be used to derive vector-valued versions of theorems on density of spaces of scalar-valued functions. This is done through the following.

COROLLARY *Let* X *be a Hausdorff space such that* $C(X)$ *separates the points of* X. *Let* $W \subset C(X;E)$ *be a subset such that* $A \otimes E \subset W$, *where* A *is the set* $\{\phi \circ f;\ \phi \in E',\ f \in W\}$. *If* A *is dense in* $C(X)$, *then* W *is dense in* $C(X;E)$.

PROOF The set $A \otimes E$ consists of all finite sums of functions of the form $x \to f(x)v$, where $f \in A$ and $v \in E$. Suppose that A is dense in $C(X)$, i.e. $\overline{A} = C(X)$. Then $C(X) \otimes E = \overline{A} \otimes E$. On the other hand, $\overline{A} \otimes E \subset \overline{A \otimes E}$. By hypothesis $A \otimes E \subset W$. Hence $C(X) \otimes E \subset \overline{W}$; by Theorem 1.15 W is dense in $C(X;E)$, and the proof is done.

If $Z \subset X$ is a closed subset, and $M \subset E$ is a closed vector subspace, then $W(Z;M) = \{f \in C(X;E); f(x) \in M$ for all $x \in Z\}$ is a closed $C(X)$-submodule. If $Z_1 \subset Z_2$, then $W(Z_2;M) \subset W(Z_1;M)$. In particular, $W(Z;M) \subset \cap\{W(\{x\};M);\ x \in Z\}$. On the other hand, if $M_1 \subset M_2$, then $W(Z;M_1) \subset W(Z;M_2)$. In particular, $W(\{x\};M) \subset \cap\{W(\{x\};H);\ M \subset H,\ \text{codim } H = 1\}$. This suggests that $W(\{x\};H)$ are maximal proper closed $C(X)$-modules, and that, in fact, each proper closed $C(X)$-module is the intersection of all maximal proper closed $C(X)$-modules that contain it.

THEOREM 1.16 *Every proper closed* $C(X)$*-module* $W \subset C(X;E)$ *is contained in some proper closed* $C(X)$*-module* V *of codimension one (hence maximal) in* $C(X;E)$. *Moreover,* W *is the intersection of all maximal proper closed* $C(X)$*-modules that contain it.*

PROOF Let $W \subset C(X;E)$ be a proper closed $C(X)$-module. Let $f \in C(X;E)$ be a function which does not belong to W. Since W is closed, by Theorem 1.11, there is $x \in X$ such that $f(x)$ does not belong to the closure of $W(x)$ in E. By the Hahn-Banach theorem, there is $\phi \in E'$ such that $\phi(f(x)) \neq 0$, while $\phi(g(x)) = 0$ for all $g \in W$. Let H be the kernel of ϕ in E, and define $V = \{g \in C(X;E);\ g(x) \in H\}$. Clearly $W \subset V$, and $f \notin V$. Since the map $T: h \to h(x)$ is a continuous linear map from $C(X;E)$ into E, and V is the kernel of $\phi \circ T$, V is a closed vector subspace of codimension one in $C(X;E)$. It remains to notice that V is a $C(X)$-module.

COROLLARY 1.17 *All maximal proper closed* $C(X)$-*modules of* $C(X;E)$ *are of the form* $\{g \in C(X;E);\ \phi(g(x)) = 0\}$ *for some* $x \in X$ *and* $\phi \in E'$.

We can generalize the above results to A-submodules, where A is any self-adjoint subalgebra. Indeed we have the following result.

THEOREM 1.18 *Let* $A \subset C(X)$ *be self-adjoint subalgebra. Every proper closed* A-*submodule* $W \subset C(X;E)$ *is contained in some proper closed* A-*submodule* V *of codimension one (hence maximal) in* $C(X;E)$. *Moreover,* W *is the intersection of all maximal proper closed* A-*submodules* V *in* $C(X;E)$ *that contain it.*

PROOF. Let $f \in C(X;E)$ be a function outside of W. Since $W = \overline{W}$, by Theorem 1.8 there is some $x \in X$ such that $f|[x]$ does not belong to the closure of $W|[x]$ in $C([x];\ E)$. (Here $[x]$ is the equivalence class (mod. A) that contains x). By the Hahn-Banach Theorem, there is $\phi \in C([x];E)'$ such that $\phi(f|[x]) \neq 0$, while $\phi(g|[x]) = 0$ for all $g \in W$. Let $V \subset C(X;E)$ be the set $\{g \in C(X;E);\ \phi(g|[x]) = 0\}$. It is clear that V is an A-module, containing W, and that $f \notin V$. Since the map $T: g \to g|[x]$ is a continuous linear map from $C(X;E)$ into $C([x];E)$, when each space carries its own compact-open topology, $V = \ker(\phi \circ T)$ is a closed vector subspace of codimension one in $C(X;E)$.

COROLLARY 1.19 *Let* $A \subset C(X)$ *be a self-adjoint subalgebra. All maximal proper closed* A-*submodules* $V \subset C(X;E)$ *are of the form*

$$V = \{g \in C(X;E);\ \phi(g|[x]) = 0\}$$

for some $x \in X$ *and* $\phi \in C([x];E)'$.

§ 7 AN EXAMPLE: A THEOREM OF RUDIN

Let E be a *real* locally convex Hausdorff space, and let F be a (complete) locally convex Hausdorff space over $\mathbb{C}$. Let $\mathcal{P}_f(E;\mathbb{R})$ be the subalgebra of $C(E;\mathbb{R})$ generated by the dual E' of E and by the constant mappings. Let $\mathcal{P}(\mathbb{C};F)$ be the vector subspace of all functions of the form

$$p(Z) = \sum_{i=0}^{n} a_i Z^i$$

where $n \in \mathbb{N}$, $a_i \in F$, $i=0,1,\ldots,n$. If $X \subset E \times \mathbb{C}$, let $\mathcal{P} \subset C(X;F)$ be the vector subspace $(\mathcal{P}_f(E;\mathbb{R}) \otimes \mathcal{P}(\mathbb{C};F))|_X$, i.e. $\mathcal{P}$ consists of the restrictions to X of finite sums of products of the form $(t,Z) \mapsto q(t)p(Z)$, where $q \in \mathcal{P}_f(E;\mathbb{R})$ and $p \in \mathcal{P}(\mathbb{C};F)$. Let $A \subset C(X;\mathbb{R})$ be the subalgebra consisting of all functions of the form $(t,Z) \in X \to q(t)$, where $q \in \mathcal{P}_f(E;\mathbb{R})$. Obviously, $\mathcal{P}$ is an A-module.

For each $t \in E$, let $K_t = \{Z \in \mathbb{C}; (t,Z) \in X\}$. For each $t \in E$, such that $K_t \neq 0$, let $X_t = \{(t,Z) \in X; Z \in K_t\}$. Since E' separates the points of E, the equivalence classes $Y \subset X$ (mod. A) are precisely the sets X_t defined above.

In the next Theorem we shall suppose that X is compact and for each $t \in E$, K_t is a compact subset of $\mathbb{C}$ which has a connected complement. In this case, we denote by $CA(X;F)$ the set of all $g \in C(X;F)$ such that, for each $t \in E$, the mapping, $g_t\colon Z \to g_t(Z) = g(t,Z)$ is holomorphic in the interior of K_t.

THEOREM 1.20 *Assume* X *is a compact Hausdorff space.* $CA(X;F)$ *is contained in the closure of* $\mathcal{P}$ *in* $C(X;F)$*, i.e. for each* $g \in CA(X;F)$*, given* $\varepsilon > 0$ *and* $r \in cs(F)$*, there exist polynomials* $q_1,\ldots,q_m \in \mathcal{P}_f(E;\mathbb{R})$ *and* $p_1,\ldots,p_m \in \mathcal{P}(\mathbb{C};F)$ *such that*

$$r(g(t,Z) - \sum_{i=1}^{m} q_i(t)\, p_i(Z)) < \varepsilon$$

for all $(t,Z) \in X$.

PROOF Let $g \in CA(X;F)$. For each $t \in E$, with $K_t \neq 0$, g_t is holomorphic in the interior of K_t. By the vector-valued version of Mergelyan's Theorem (see Bierstedt [5], and Briem, Lauersen, and Pedersen [9]), the function g_t belongs to the closure of

$\mathcal{P}(\mathbb{C};F)|K_t$ in $C(K_t;F)$. Therefore, given $\varepsilon > 0$ and $r \in cs(F)$, there exists $p \in \mathcal{P}(\mathbb{C};F)$ such that $r(g_t(Z) - p(Z)) < \varepsilon$ for all $Z \in K_t$. Let $f = (1 \otimes p)|X$. Then $f \in \mathcal{P}$ and $r(g(t,Z) - f(t,Z)) < \varepsilon$ for all $(t,Z) \in X_t$. It follows that $g \in L_A(\mathcal{P})$. By Theorem 1.5 g belongs to the closure of $\mathcal{P}$, QED.

When $E = \mathbb{R}^n$ and $F = \mathbb{C}$, the above result was proved by Rudin ([54], Th.4). His proof is a partition of unity argument combined with Mergelyan's Theorem. In his book [56] Rudin derives it from Bishop's theorem (see the proof of Theorem 5.8, [56]). When $E = \mathbb{R}^I$, where I is an arbitrary set of indices, and $F = \mathbb{C}$, Theorem 1 was proved by Chalice [13]. His very short proof is based on de Branges lemma that says that for any extreme point μ of the unit ball of $A^\perp$, the support of μ is a set of antisymmetry for A, if A is any function algebra. This lemma is an essential step in Glicksberg's proof [27] of Bishop's Theorem. Our next result is a corollary of Theorem 1.20 above. Again, when $I = \{1,\dots,n\}$, $K = [0,1] \subset \mathbb{R}$, it is due to Rudin ([54], Theorem 3); and when I is an arbitrary index set, and $K \subset \mathbb{R}$ is an arbitrary compact subset of the line, it is due to Chalice ([13], Theorem 2). Chalice presents a direct proof, very short, based again on de Branges lemma and Mergelyan's Theorem.

COROLLARY 1.21 *Let* K *be a compact subset of* $\mathbb{R}$. *If* $f \in C(K;\mathbb{C})$ *and* $u_i \in C(K;\mathbb{R})$, $i \in I$, *are such that* f *and the* u_i *separate the points of* K, *then the function algebra* A *on* K *generated by* f *and the* u_i *is* $C(K;\mathbb{C})$.

PROOF Consider the real locally convex Hausdorff space $E = \mathbb{R}^I$. The mapping ϕ defined on K by $x \to ((u_i(x))_{i\in I}, f(x))$ is a homeomorphism onto a compact subset $X \subset E \times \mathbb{C}$. For each $t \in E$, let $K_t = \{Z \in \mathbb{C};\ (t,Z) \in X\}$. Each such compact subset has a connected complement in $\mathbb{C}$, and the interior of K_t is empty. Therefore, $CA(X;\mathbb{C}) = C(X;\mathbb{C})$.

Let $h \in C(K;\mathbb{C})$. Then $g = h \circ \phi^{-1}$ belongs to $C(X;\mathbb{C})$. For each $\varepsilon > 0$ there exist $q_1,\dots q_m \in \mathcal{P}_f(\mathbb{R}^I;\mathbb{R})$ and $p_1,\dots,p_m \in \mathcal{P}(\mathbb{C};\mathbb{C})$ such that

$$\left|\sum_{j=1}^{m} q_j(t)\, p_j(Z) - g(t,Z)\right| < \varepsilon$$

for all $(t,Z) \in X$. Now each $(t,Z) \in X$ is of the form $\phi(x)$ for some $x \in K$. Therefore $g(t,Z) = h(x)$. On the other hand each $q_j \in \mathcal{P}_f(\mathbb{R}^I;\mathbb{R})$ is a finite sum of homogeneous polynomials of the form $\pm\phi^k$ $(k \in \{0,1,2,\ldots,n,\ldots\})$, with $\phi \in E' = \bigoplus_{i \in I} \mathbb{R}$. For each such ϕ, there exists a *finite* set $F \subset I$ such that

$$\phi((t_i)_{i\in I}) = \sum_{i\in F} a_i t_i$$

where $a_i \in \mathbb{R}$ for all $i \in F$. Hence

$$\left[\phi((u_i(x))_{i\in I})\right]^k = \left[\sum_{i\in F} a_i u_i(x)\right]^k$$

for all $x \in K$. Consequently, there exists a finite sum of homogeneous polynomials $Q_F^j \in \mathcal{P}^k(\mathbb{R}^{|F|};\mathbb{R})$, $F \subset I$ finite, $|F|$ = cardinal of F, such that

$$q_j(t) = \sum_{\text{finite}} Q_F^j(u_{\alpha_1}(x),\ldots,u_{\alpha_{|F|}}(x))$$

for all $t = (u_i(x))_{i\in I}$, where $F=\{\alpha_1,\ldots,\alpha_{|F|}\}$.

Now the function

$$w = \sum_{j=1}^{m} \left(\sum_{\text{finite}} Q_F^j(u_{\alpha_1},\ldots,u_{\alpha_{|F|}})\right) p_j(f)$$

belongs to A, and $w(x) = \sum_{j=1}^{m} q_j(t)p_j(Z)$ for all $x \in K$, if $(t,Z) = \phi(x)$. Therefore $|w(x) - h(x)| < \varepsilon$, for all $x \in K$. Hence the algebra A is dense and closed in $C(K;\mathbb{C})$, i.e., $A = C(K;\mathbb{C})$, as desired.

REMARK. We shall present a proof of the vector-valued Mergelyan's Theorem in Chapter 8. A simple proof, which is available when the space E has the *approximation property*, is presented in Chapter 4. A locally convex space E has the approximation property, if given $K \subset E$ compact $p \in cs(E)$, $\varepsilon > 0$, there is a continuous linear mapping u of finite rank from E to E, i.e. an element $u \in E' \otimes E$, such that $p(x - u(x)) < \varepsilon$ for all $x \in K$.

§ 8 BISHOP'S THEOREM

In the theorems of this paragraph X denotes a *compact* Hausdorff space and E a seminormed space, with seminorm

$t \to ||t||$. Let $A \subset C(X;\mathbb{R})$ be a subalgebra and $W \subset C(X;E)$ an A-module. For each $x \in X$, we denote by $[x]$ be equivalence class (mod. A) that contains x, and W_x denotes the vector space $W|[x]$ contained in $C([x];E)$. Moreover, we write $||f|| = \sup\{||f(t)||; t \in X\}$ and $||f|_{[x]}|| = \sup\{||f(t)||;\ t \in [x]\}$. Using this notation, the following stronger form of Theorem 1.5 can be proved (See Buck [12], Theorem 2, pg. 87; Glicksberg [26], pg. 419).

THEOREM 1.22 *For every* $f \in C(X;E)$,

$$\inf_{g \in W} ||f-g|| = \sup_{x \in X} \inf_{g \in W} ||f|_{[x]} - g|_{[y]}|| .$$

PROOF Let $d = \inf\{||f-g||; g \in W\}$; and for each $x \in X$, let $\lambda(x) = \inf\{||f|_{[x]} - g|_{[x]}||; g \in W\}$ and $\lambda = \sup\{\lambda(x);\ x \in X\}$.

Since $[x] \subset X$, $\lambda(x) \leq d$ for all $x \in X$. Hence $\lambda \leq d$.

To prove the reverse inequality, let $0 < \varepsilon < 1$. Without loss of generality we may assume that A contains the constants. For each $x \in X$, there exists $w_x \in W$ such that

$$||f(t) - w_x(t)|| < \lambda + \varepsilon/3 \text{ for all } t \in [x].$$

Let $U_x \subset X$ be the open subset $\{t \in X; ||f(t) - w_x(t)|| < \lambda + \varepsilon/3\}$. Then $[x] \subset U_x$. Proceeding exactly as in the proof of Theorem 1.5 (recall that we have assumed $A \subset C(X;\mathbb{R})$) we find $g \in W$ such that $||f(x) - g(x)|| < \lambda + \varepsilon$ for all $x \in X$, i.e. $||f-g|| < \lambda + \varepsilon$. Thus $d < \lambda + \varepsilon$. Since $\varepsilon > 0$ was arbitrary, $d \leq \lambda$. This completes the proof of the Theorem.

REMARK In the proof of the inequality $\lambda \leq d$, we did not use the fact that $A \subset C(X;\mathbb{R})$. In proving the reverse inequality $d \leq \lambda$, if $A \subset C(X;\mathbb{C})$ is a *self-adjoint* subalgebra, we may use Lemma 1.7, §5, to substitute $B = \text{Re } A = \{\text{Re } f;\ f \in A\}$ for A in the proof. Hence the following result is true.

THEOREM 1.23 *Let* $A \subset C(X;\mathbb{C})$ *be a self-adjoint subalgebra, and* $W \subset C(X;E)$ *an A-module. For each* $f \in C(X;E)$,

$$\inf_{g \in W} ||f-g|| = \sup_{x \in X} \inf_{g \in W} ||f|_{[x]} - g|_{[x]}||.$$

COROLLARY 1.24 (Buck [12]) *Let* $W \subset C(X;E)$ *be a* $C(X)$*-module. For each* $f \in C(X;E)$ *we have*

$$\inf_{g \in W} ||f-g|| = \sup_{x \in X} \inf_{g \in W} ||f(x)-g(x)||.$$

An even stronger form of Theorems 1.22 and 1.23 is available. To prove this we notice the following

LEMMA 1.25 *(Machado and Prolla [39], pg. 126) Let* X *and* Y *be compact Hausdorff spaces and* π *a continuous mapping from* X *onto* Y. *For each upper semicontinuous function* $g: X \to \mathbb{R}$ *define* $h: Y \to \mathbb{R}$ *by*

$$h(y) = \sup \{g(x);\ x \in \pi^{-1}(y)\}$$

for all $y \in Y$. *Then* h *is upper semicontinuous on* Y.

PROOF For each $y \in Y$, the set $\{y\}$ is closed in Y, therefore $\pi^{-1}(y)$ is compact in X. Hence there is an $a \in \pi^{-1}(y)$ such that $h(y) = g(a) = \sup \{g(x);\ x \in \pi^{-1}(y)\}$, because g is upper semicontinuous. So h is well defined from Y to $\mathbb{R}$. Let $r \in \mathbb{R}$. The set $\{x \in X;\ g(x) \geq r\}$ is closed, whence compact in X. Call it X_r. Since π is continuous, $\pi(X_r)$ is compact in Y. We claim that $\pi(X_r) = \{y \in Y;\ h(y) \geq r\}$, which proves that h is upper semicontinuous. Indeed, if $y \in \pi(X_r)$, then $y = \pi(x)$ for some $x \in X_r$, and then $h(y) \geq g(x) \geq r$. Conversely, if $y \notin \pi(X_r)$ and $t \in \pi^{-1}(y)$, then $g(t) < r$; it follows that $h(y) < r$, because there is some point $t_0 \in \pi^{-1}(y)$ such that $h(y) = g(t_0)$. That completes the proof.

THEOREM 1.26 *Let* $A \subset C(X;\mathbb{R})$ *be a subalgebra and let* $W \subset C(X;E)$ *be an* A*-module. Let* $f \in C(X;E)$. *There exists a point* $x \in X$ *such that*

$$\inf_{g \in W} ||f-g|| = \inf_{g \in W} ||f|_{[x]} - g|_{[x]}||$$

PROOF Let Y be the compact Hausdorff space that is the quotient of X by the equivalence relation defined by A, and let

$\pi : X \to Y$ be the quotient map. Then, by Lemma 1.25 the function

$$[x] \to \| f|_{[x]} - g|_{[x]} \|$$

is upper semicontinuous on Y, for each $g \in W$. Hence

$$[x] \to \inf_{g \in W} \| f|_{[x]} - g|_{[x]} \|.$$

is upper semicontinuous on Y too, and therefore attains its supremum. By Theorem 1.22 this supremum is $d = \inf\{\|f-g\|; g \in W\}$.

We will prove now a remarkable fact discovered by S. Machado [37]. Namely that Theorem 1.26 above implies Bishop's Theorem. To do so we will use Bishop's original description of the partition of X into antisymmetric sets (See [8]).

If X is a Hausdorff space and $A \subset C(X;\mathbb{K})$ is a subalgebra, let $\mathfrak{S}$ be the class of all ordinal numbers whose cardinal numbers are less or equal to $2^{|X|}$, where $|X|$ is the cardinal number of X. For each $\sigma \in \mathfrak{S}$, we define by transfinite induction a closed, pair-wise disjoint covering of X, denoted by P_σ.

For $\sigma = 1$, we define $P_1 = \{X\}$.

Assume that P_τ has been defined for all ordinals $\tau < \sigma$. We consider two cases.

(a) If $\sigma = \tau+1$, for some $\tau \in \mathfrak{S}$, we define P_σ as follows. Let $T \subset X$ be an element of P_τ. Let $A_T = \{f \in A; f|_T$ is real$\}$. Then $A_T|T \subset C(T;\mathbb{R})$ and we consider the partition of T into equivalence classes (mod. A_T). The partition P_σ is then defined as the collection of all such equivalence classes when T ranges over P_τ.

(b) If σ has no predecessor, i.e. σ is a limit ordinal, define $x \equiv y$ for $x,y \in X$, if, and only if, x and y belong to the same equivalence class of P_τ for all $\tau < \sigma$.

This defines P_σ for all $\sigma \in \mathfrak{S}$, and P_σ is a refinement of P_τ whenever $\sigma > \tau$. Bishop's argument that there exists an ordinal $\rho \in \mathfrak{S}$ such that each element $S \in P_\rho$ is antisymmetric for A is as follows. We recall that a subset $S \subset X$ is *antisymmetric for* A if, for any $f \in A$, the restriction $f|S$ is real-valued implies that $f|_S$ is constant.

Assume that $P_{\sigma+1}$ is a proper refinement of P_σ for all $\sigma \in \mathfrak{S}$. Then $P_{\sigma+1}$ contains a set not in P_τ for all $\tau < \sigma + 1$. Therefore the cardinal number of subsets of X is $\geq |\mathfrak{S}|$. This contradicts the definition of $\mathfrak{S}$. Hence there exists an ordinal $\rho \in \mathfrak{S}$, such that $P_\rho = P_{\rho+1}$. Whence $P_\rho = \mathcal{K}_A$, where $\mathcal{K}_A$ denotes the closed, pair-wise disjoint, partition of X into maximal antisymmetric sets for A.

THEOREM 1.27 (*Bishop* [8]; *Glicksberg* [26]) *Let* X *be a compact Hausdorff space and let* $A \subset C(X;\mathbb{K})$ *be a real subalgebra. Let* $W \subset C(X;\mathbb{K})$ *be an* A*-module. For each* $f \in C(X;\mathbb{K})$, f *belongs to the closure of* W *if, and only if,* $f|S$ *belongs to the closure of* $W|S$ *in* $C(S;\mathbb{K})$, *for all* $S \in \mathcal{K}_A$.

PROOF This is an immediate corollary of the following stronger for of Theorem 1.27.

THEOREM 1.28 (*Machado* [37]) *Let* X *be a compact Hausdorff space and let* E *be a seminormed space. Let* $A \subset C(X;\mathbb{K})$ *be a real subalgebra and* $W \subset C(X;E)$ *an* A*-module. Let* $f \in C(X;E)$. *For each* $\sigma \in \mathfrak{S}$, *there is* $S_\sigma \in P_\sigma$ *such that*

(a) $S_\sigma \subset S_\tau$ *for all* $\tau < \sigma$, $\tau \in \mathfrak{S}$;

(b) $\inf_{g \in W} ||f-g|| = \inf_{g \in W} ||f|S_\sigma - g|S_\sigma||$.

PROOF By a real subalgebra $A \subset C(X;\mathbb{C})$ we mean that the algebra A is an $\mathbb{R}$-algebra.

Let $\sigma \in \mathfrak{S}$.

Assume that, given $f \in C(X;E)$, a set $S_\tau \in P_\tau$ with properties (a) and (b) has been found for all $\tau < \sigma$.

1st CASE $\sigma = \tau + 1$, with $\tau \in \mathfrak{S}$. By the induction hypothesis, there is $S_\tau \in P_\tau$ such that $S_\tau \subset S_\mu$ for all $\mu \in \mathfrak{S}$, $\mu < \tau$ and

$$\inf_{g \in W} ||f-g|| = \inf_{g \in W} ||f|S_\tau - g|S_\tau||.$$

Let $A_\tau \subset A$ be the subalgebra of all $h \in A$ such that $h|S_\tau$ is real. By Theorem 1.26 applied to the algebra $A_\tau|S_\tau$ and the module $W|S_\tau$ (over $A_\tau|S_\tau$) there is a set $S_\sigma \in P_\sigma = P_{\tau+1}$ such that

$$\inf_{g\in W} ||f|S_\tau - g|S_\tau|| = \inf_{g\in W} ||f|S_\sigma - g|S_\sigma||$$

On the other hand $S_\sigma \subset S_\tau$ by construction. This proves (a) and (b) in this case.

2nd CASE. The ordinal σ has no predecessor. Define $S_\sigma = \bigcap_{\tau<\sigma} S_\tau$. Then $S_\sigma \in P_\sigma$ and $S_\sigma \subset S_\tau$ for all $\tau < \sigma$, $\tau \in \mathfrak{S}$. To prove (b) assume by contradiction that

$$\inf_{g\in W} ||f|S_\sigma - g|S_\sigma|| < d$$

where $d = \inf \{||f-g||;\ g \in W\}$.

(The case $d = 0$ is trivial). It follows that there is some $g \in W$ such that

$$||f|S_\sigma - g|S_\sigma|| < d$$

Let $U = \{t \in X; ||f(t) - g(t)|| < d\}$.

Then $S_\sigma \subset U$, and U is open. By compactness of $X \setminus U$ there are finitely many indices $\tau_1 < \tau_2 < \ldots < \tau_n < \sigma$ such that

$$X \setminus U \subset (X \setminus S_{\tau_1}) \cup \ldots \cup (X \setminus S_{\tau_n}).$$

Since $S_{\tau_1} \subset S_{\tau_2} \subset \ldots \subset S_{\tau_n}$, it follows that $S_{\tau_n} \subset U$, which contradicts (b) for $\tau_n \in \mathfrak{S}$, $\tau_n < \sigma$. This proves (b) for the ordinal σ, and ends the proof of Theorem 1.28.

REMARK In his proof of Theorem 1.28 S.Machado [37] uses Zorn's Lemma instead of the above transfinite argument. The idea of using Zorn's Lemma can be applied to give a direct proof of the bounded case of the Bernstein-Nachbin approximation problem (see Machado and Prolla [41]).

Notice that P_2 is the partition of X introduced by Shilov. Indeed P_2 is the collection of all equivalence classes of X modulo the algebra $\{f \in A;\ f \text{ is real on } X\}$. This coarser of all proper partitions of X can be used sometimes instead of $\mathcal{K}_A$. It then follows from Theorem 1.28 that there exists some $S \in P_2$(once $f \in C(X;E)$ is given) such that

$$\mathrm{dist}(f,W) = \mathrm{dist}(f|S\ ;W|S\ ).$$

This formula in turn implies the following.

COROLLARY 1.29 *Let* A *and* W *be as in Theorem 1.28.Let* $f \in C(X;E)$. *Then* f *belongs to the uniform closure of* W *if, and only if,* $f|S$ *belongs to the uniform closure of* $W|S$ *in* $C(S;E)$, *for all* $S \subset X$ *which is an equivalence class modulo the algebra*$\{g \in A;\ g$ is real on $X\}$.

§ 9 VECTOR FIBRATIONS

A *vector fibration* is a pair $(X;(E_x;\ x \in X))$, where X is a Hausdorff topological space and $(E_x;\ x \in X)$ is a family of vector-spaces over the same scalar field $\mathbb{K}$.

The product set $\pi(E_x;\ x \in X)$ is always provided with the structure of a product vector-space over $\mathbb{K}$; it is called the vector space of all *cross-sections* of the vector fibration $(X;(E_x;\ x \in X))$. A *vector space of cross-sections* is then any one of the vector subspaces of $\pi(E_x;\ x \in X)$.

A vector space W of cross-sections is said to be a *module over a subalgebra* $A \subset C(X;\mathbb{K})$, or an A-*module*, if $aw \in W$ for any $a \in A$ and $w \in W$, where aw is the cross-section $(a(x)w(x);\ x \in X)$ if $(w(x);\ x \in X) = w$.

Any family $\nu = (\nu_x;\ x \in X)$ such that ν_x is a seminorm on E_x for each $x \in X$ is called a *weight* of the vector fibration $(X;(E_x;\ x \in X))$.

We shall restrict our attention to vector fibrations and vector spaces L of cross-sections satisfyng the following conditions:

(1) X is compact;

(2) each E_x is a normed space, whose norm we denote by $t \to ||t||$;

(3) if L is a vector space of cross sections, for each $f \in L$ the function $x \to ||f(x)||$ is upper semicontinuous on X.

In this case we say that L is an *upper semicontinuous vector space of cross-sections*, and endow it with the topology of the norm

$$||f|| = \sup \{||f(x)||;\ x \in X\}.$$

THEOREM 1.30 ([15], *Lemma* 4) *Let* L *be an upper semicontinuous vector-space of cross-sections which is a* C(X)*-module. For every* C(X)*-submodule* $W \subset L$, *we have*

$$\inf_{g \in W} ||f-g|| = \sup_{x \in X} \inf_{g \in W} ||f(x)-g(x)||$$

for every $f \in L$.

PROOF The proof is entirely similar to that of Theorem 1.22, § 8, only it is much simpler. Call $\rho = \inf_{g \in W} ||f-g||$. For each $x \in X$, let $\lambda(x) = \inf_{g \in W} ||f(x)-g(x)||$ and put $\lambda = \sup\{\lambda(x); x \in X\}$. Clearly $\lambda(x) \leq \rho$, for each $x \in X$. Hence $\lambda \leq \rho$.

To prove the reverse inequality, let $0 < \varepsilon$. For each $x \in X$, there exists $w_x \in W$ such that $||f(x)-w_x(x)|| < \lambda + \varepsilon$. Since $f - w_x \in L$, by condition (3), $U_x = \{t \in X;\ ||f(t)-w_x(t)|| < \lambda + \varepsilon\}$ is an open set containing x. By compacteness of X, there exist $x_1,\ldots,x_n \in X$ such that $X = U_1 \cup \ldots \cup U_n$, where for each $i = 1,2,\ldots,n$, U_i denotes U_x for $x = x_i$. Let $g_1,\ldots,g_n \in C(X)$ be a continuous partition of the unity subordinate to the covering $U_1,\ldots,U_n$. Let $g = \sum_{i=1}^{n} g_i w_i$, where $w_i = w_x$ for $x = x_i$ $(i = 1,2,\ldots n)$. Since W is a C(X)-submodule, $g \in W$. We claim that, for each $x \in X$,

$$g_i(x)||f(x)-w_i(x)|| \leq g_i(x)(\lambda + \varepsilon),$$

for all $i = 1,2,\ldots,n$. Indeed, if $x \notin U_i$ then $g_i(x) = 0$ and both sides are zero. If $x \in U_i$, then $||f(x)- w_i(x)||< \lambda + \varepsilon$, and $g_i(x) \geq 0$. Hence

$$||f(x)- g(x)|| = ||\sum_{i=1}^{n} g_i(x)(f(x)- w_i(x))||$$

$$\leq \sum_{i=1}^{n} g_i(x) \; ||f(x)- w_i(x)|| \leq \sum_{i=1}^{n} g_i(x)(\lambda+\varepsilon)$$

$$= \lambda + \varepsilon, \text{ for all } x \in X.$$

Therefore $||f-g|| \leq \lambda + \varepsilon$, and a fortiori, $\rho \leq \lambda + \varepsilon$. Since $\varepsilon > 0$ was arbitrary, $\rho \leq \lambda$. This concludes the proof of Theorem 1.30.

REMARK. Theorem 1.30 is a generalization of Corollary 1.24, §8. To see this, just consider the vector fibration $E_x = E$ for all $x \in X$, where E is a fixed normed space. And then take $L=C(X;E)$. Obviously, if $W_x = \overline{W(x)}$ in E, then

$$\inf_{u \in W_x} ||f(x)- u|| = \inf_{g \in W} ||f(x)- g(x)||.$$

§ 10 EXTREME FONCTIONALS

Let S be a subset of a vector space E. A point $x \in S$ is called an *extreme point* of S if x is not an internal point of any segment of line in S, i.e., if $a,b \in S$ and $0 < t < 1$, then $ta + (1-t)b = x$ implies that $a = b = x$. The set of extreme points of the set S is denoted by $E(S)$.

Let us recall, for use in Lemma 1.33 below, that the *convex hull* of a set $S \subset E$ is the smallest convex set in E that contains the set S, and it is denoted by $co(S)$. When E is a topological vector space, the *closed convex hull* of S, denoted by $\overline{co}(S)$, is the closure of its convex hull.

We shall be interested in this section in characterizing $E(S)$, when S is a subset of the dual of some locally convex space L of functions, or more generally, of cross-sections. Let $L = C(X)$, and let S be the unit ball of $E = L'$; the Arens-Kelley Theorem ([2], or Dunford and Schwartz [20], pg.441) says that $E(S)$ are the evaluations at points of X multiplied by scalars of absolute value one. If $L = C(X;E)$, where E is a Banach space, and S = unit ball of L', then Singer characterized $E(S)$ as follows. Let B_E' be the unit ball of E'. For each $x \in X$, let $\delta_x : E' \to C(X;E)'$ be defined by

$$\delta_x(\phi)(f) = \phi(f(x))$$

for all $f \in C(X;E)$ and $\phi \in E'$. Then

$$\text{(a)} \quad E(S) = \cup\{\delta_x(E(B_E')); \ x \in X\}.$$

(See Singer [60]). This result was generalized to the space $C_o(X;E)$ of all $f \in C(X;E)$ which vanish at infinity on a *locally compact* Hausdorff space X, by Brosowski, Deutsch and Morris. (See [10], Lemma 3.3), and by Ströbele [63].

The Arens-Kelley theorem was extended by R.C. Buck [12] in the following direction. If $L = C(X;E)$, and $M \subset L$ is a $C(X)$-module, $M \neq \{0\}$, let S be the convex set $M^\perp \cap B'$, where B' is the unit ball of L'. For each $x \in X$, let M_x be the closure in E of $M(x) = \{f(x); \ f \in M\}$. Then

$$\text{(b)} \quad E(M^\perp \cap B') = \cup\{\delta_x(E(M_x^\perp \cap B_E'));x \in X, \ M_x \neq E\}.$$

Previously, Ströbele had proved (see [63]) the inclusion $\supset$.

Independently, Cunningham and Roy [15] had proved (b) for $C(X)$-modules $M \subset L$, where L is a vector space of cross-sections satisfying the hypothesis of Theorem 1.30. We shall present this more general result because we shall reduce the case of a general A-module $W \subset C(X;E)$, where A is any self-adjoint subalgebra, not necessarily separating on X, to the case of $C(X)$-modules of cross-sections, through a quotient construction.

We start with Singer's Theorem for cross-sections.

Let B' be the unit ball of L', and for each $x \in X$, let B_x' be the unit ball of E_x'. The mapping $\delta_x: E_x' \to L'$ defined by

$$\delta_x(\phi)(f) = \phi(f(x))$$

for all $\phi \in E_x'$ and $f \in L$, is then a continuous linear mapping of norm ≤ 1, and therefore maps B_x' into B'.

Notice also that δ_x is one-to-one if

$L_x = \{f(x);\ f \in L\} = E_x$. This leads to the following:

DEFINITION 1.31 *A vector space* L *of cross-sections is said to be essential if* $L_x = \{f(x);\ f \in L\} = E_x$, *for all* $x \in X$.

Clearly, the image of E_x' by δ_x is contained in the set $\{\psi \in L'; f(x) = 0 \Rightarrow \psi(f) = 0,\ \forall\ f \in L\}$. Call it A_x. We claim the following.

LEMMA 1.32 *Let* L *be an essential* C(X)*-module of upper semi-continuous cross-sections over* X. *The mapping* δ_x *is a linear isometry of* E_x' *onto* A_x *for each* $x \in X$.

PROOF We saw already that δ_x is a continuous linear mapping of norm ≤ 1, from E_x' into A_x.

Let $\psi \in A_x$. For each $v \in E_x$, choose $f \in L$ such that $f(x) = v$ and put $\varepsilon_x(\psi)(v) = \psi(f)$. Since $\psi \in A_x$, $\varepsilon_x(\psi)$ is well defined on E_x, and it is clearly a linear functional. We claim that $\varepsilon_x(\psi) \in E_x'$. Let $\varepsilon > 0$. Let $v \neq 0$ be given in E_x. Choose $f \in L$ with $f(x) = v$. By the upper semicontinuity hypothesis, there is a neighborhood U of x in X such that $||f(t)|| < (1 + \varepsilon)||f(x)||$, for all $t \in U$. Let $g \in C(X)$ satisfy $0 \leq g \leq 1$, $g(x) = 1$ and $g(t) = 0$ for all $t \notin U$. Since L is a C(X)-module, $gf \in L$, and moreover

$$||gf|| < (1+\varepsilon)\ ||f(x)||. \text{ Now } (gf)(x) = v \text{ and}$$

$$|\varepsilon_x(\psi)(v)| = |\psi(gf)| \leq ||\psi||.||gf||$$

$$< ||\psi||.(1+\varepsilon).||f(x)|| = ||\psi||.(1+\varepsilon).||v||.$$

Hence $\varepsilon_x(\psi) \in E_x'$ and $||\varepsilon_x(\psi)|| \leq ||\psi||$ for all $\psi \in A_x$. Therefore $||\varepsilon_x|| \leq 1$. Since δ_x and ε_x are inverses of each other, and $||\delta_x|| \leq 1$, we see that δ_x and ε_x are linear isometries.

LEMMA 1.33 *Let* L *be an essential upper semicontinuous vector-space of cross-sections over* X. *Let* $Q = \cup\{\delta_x(B_x');\ x \in X\}$. *Then*
(a) Q *is weak*-closed;*
(b) $\overline{\text{co}}\ (Q) = B'$.

PROOF (a) Suppose $\psi \in L'$ is the weak*-limit of a net $\{\psi_\alpha\}$ in

Q. Each ψ_α is of the form $\psi_\alpha = \delta_{x_\alpha}(\phi_\alpha)$, where $x_\alpha \in X$ and $\phi_\alpha \in B'_{x_\alpha}$. Since X is compact we may assume that $\{x_\alpha\}$ converges to some point $x \in X$. Then, for any $f \in L$ we have

$$|\psi(f)| = \lim |\psi_\alpha(f)| = \lim |\phi_\alpha(f(x_\alpha))|$$
$$\leq \limsup ||\phi_\alpha|| \cdot ||f(x_\alpha)|| \leq ||f(x)||.$$

Hence $\psi(f) = 0$, if $f(x) = 0$, i.e. $\psi \in A_x$. By Lemma 1.32 there is some $\phi \in E'_x$ such that $\delta_x(\phi) = \psi$, or equivalently, $\phi = \varepsilon_x(\psi)$. The above inequality shows that $||\psi|| = ||\phi|| \leq 1$, i.e. $\phi \in B'_x$. Therefore $\psi = \delta_x(\phi) \in Q$, as desired.

(b) Since $\delta_x(B'_x) \subset B'$, $Q \subset B'$. On the other hand B' is convex, and weak*-compact by Alaoglu's Theorem; hence the inclusion $\overline{co}(Q) \subset B'$ obtains. (By $\overline{co}(Q)$ we mean the weak*-closed convex balanced hull of Q). Let $f \in Q^o$ = the polar of Q in L. For each $x \in X$, by the Hahn-Banach Theorem there is $\phi \in B'_x$ such that $\phi(f(x)) = ||f(x)||$. Then $\delta_x(\phi) \in Q$. Hence $||f(x)|| = |\phi(f(x))| = |\delta_x(\phi)(f)| \leq 1$. This shows that $||f|| \leq 1$, i.e. $Q^o \subset B$ = unit ball of L. Taking polars we obtain $B' \subset Q^{oo}$. By the bipolar Theorem, $Q^{oo} = \overline{co}(Q)$, and that ends the proof.

THEOREM 1.34 *(Cunningham and Roy [15]). Let* L *be an essential* C(X)*-module of upper semicontinuous cross-sections over* X. *Then*

$$E(B') = \cup\{\delta_x(E(B'_x));\ x \in X,\ E_x \neq \{0\}\}$$

PROOF By Lemma 5, Dunford and Schwartz [20], pg. 440, $E(B') \subset Q$. Therefore any $\psi \in E(B')$ is of the form $\delta_x(\phi)$ for some $\phi \in B'_x$, $x \in X$. But in this case $\phi \in E(B'_x)$, because $\psi = \delta_x(\phi) \in E(B')$ and δ_x is linear and one-to-one.

Conversely, let $\psi \in \delta_x(E(B'_x))$ for some $x \in X$, $E_x \neq \{0\}$. Let $\phi \in E(B'_x)$ be such that $\psi = \delta_x(\phi)$. Notice that $\psi \in A_x \cap B'$ and $\phi = \varepsilon_x(\psi)$. But in this case $\psi \in E(A_x \cap B')$, because $\phi = \varepsilon_x(\psi) \in E(B'_x)$ and ε_x is an isometry. To complete the proof we must establish the following

CLAIM 1.35 $\psi \in E(B')$.

PROOF Assume, by contradiction, that $\psi \notin E(B')$. Then

$\psi = (\psi_1 + \psi_2)/2$ for some $\psi_1, \psi_2 \in B'$ and $\psi_1 \neq \psi_2$. Let $f \in L$ with $f(x) = 0$, $||f|| \leq 1$, and $0 < \varepsilon < 1$ be given. Then $U = \{t \in X; ||f(t)|| < \varepsilon\}$ is open and contains x. Let $g \in C(X)$ with $0 \leq g \leq 1$, $g(x) = 1$ and $g(t) = 0$ for $t \notin U$ be chosen Since $||\phi|| = 1$, there is $v \in E_x$ such that $||v|| < 1$, $\phi(v)$ is real and $\phi(v) > 1 - \varepsilon$. Choose $h \in L$ with $||h|| \leq 1$, $h(x) = v$. Let $m = gh$. Then $||gh|| \leq 1$, and $\phi(m(x)) = \phi(h(x)) = \phi(v) > 1 - \varepsilon$. On the other hand, $m(t) = 0$ for $t \notin U$ and $||f(t)|| < \varepsilon$ for $t \in U$, imply $||f(t) + m(t)|| < 1 + \varepsilon$ for all $t \in X$. Hence $||f + m|| \leq 1 + \varepsilon$. Now $|\psi_1(m)| \leq 1$ and $|\psi_2(m)| \leq 1$. Thus $\psi_1(m)$ and $\psi_2(m)$ are complex numbers in the unit disk whose mid-point $\psi(m) = \delta_x(\phi)(m) = \phi(m(x))$ is real and $> 1 - \varepsilon$. Hence

$$(1) \quad |\psi_1(m) - \psi_2(m)| < 2\sqrt{1-(1-\varepsilon)^2} < 4\sqrt{\varepsilon}$$

On the other hand, $|\psi_1(f+m)| \leq 1+\varepsilon$ and $|\psi_2(f+m)| \leq 1 + \varepsilon$. Thus $\psi_1(f+m)$ and $\psi_2(f+m)$ are complex numbers in the disk of radius $1+\varepsilon$ whose mid-point $\psi(f+m) = \psi(f) + \psi(m) = \psi(m)$ (because $f(x) = 0$ and $\psi \in A_x$) is real and $> 1-\varepsilon$. Hence

$$(2) \quad |\psi_1(f+m) - \psi_2(f+m)| < 4\sqrt{\varepsilon}$$

Combining (1) and (2) we get

$$|\psi_1(f) - \psi_2(f)| < 8\sqrt{\varepsilon},$$

whence $\psi_1(f) - \psi_2(f) = 0$. This proves that $\psi_1 - \psi_2 \in A_x$. Since $\psi_1 + \psi_2 = 2\psi \in A_x$ too, ψ_1 and ψ_2 belong to $A_x \cap B'$, which contradicts the fact that $\psi \in E(A_x \cap B')$.

COROLLARY 1.36 *(Singer) Let* X *be a compact Hausdorff space and let* E *be a normed space. Let* B' *be the unit ball of the dual of* C(X;E). *Then*

$$E(B') = \cup \{\delta_x(E(B'_E)); x \in X\},$$

where B'_E *is the unit ball of the dual* E' *of* E, *and for each* $x \in X$, *the mapping* $\delta_x\colon E' \to C(X;E)'$ *is defined by*

$$\delta_x(\phi)(f) = \phi(f(x))$$

for all $f \in C(X;E)$ *and* $\phi \in E'$.

COROLLARY 1.37 *(Arens and Kelley [2]) Let* X *be a compact Hausdorff space. Let* B' *be the unit ball of the dual of* $C(X;\mathbb{K})$. *Then*

$$E(B') = \{\lambda\delta_x;\ x \in X,\ \lambda \in \mathbb{K},\ |\lambda| = 1\},$$

where $\lambda\delta_x : C(X;\mathbb{K}) \to \mathbb{K}$ *is defined by*

$$\lambda\delta_x(f) = \lambda\cdot f(x)$$

for all $f \in C(X;\mathbb{K})$.

We come now to the case of general $C(X)$-modules $M \subset L$. Since the case $\overline{M} = L$ is covered by Theorem 1.34, we shall assume $\overline{M} \neq L$. For any $x \in X$, we define $M_x = \{f(x);\ f \in M\} \subset E_x$. We can identify $(E_x/\overline{M}_x)'$ with $M_x^{\perp}$, the unit ball being $M_x^{\perp} \cap B_x'$, respectively $(L/\overline{M})'$ with $M^{\perp}$, the unit ball being $M^{\perp} \cap B'$. We have then

THEOREM 1.38 *(Cunningham and Roy [15])*

$$E(M^{\perp} \cap B') = \cup\{\delta_x(E(M_x^{\perp} \cap B_x'));\ x \in X,\ \overline{M}_x \neq E_x\}.$$

Before proceeding to the proof, we notice a corollary for $L = C(X;E)$, $M \subset L$ a $C(X)$-module, B' the unit ball of L', for each $x \in X$, $M_x = \{f(x);\ f \in M\} \subset E$, and B_E' unit ball of E'.

COROLLARY 1.39 *(Buck [12])*

$$E(M^{\perp} \cap B') = \cup\{\delta_x(E(M_x^{\perp} \cap B_E'));\ x \in X,\ \overline{M}_x \neq E\}.$$

PROOF OF THEOREM 1.38. We will identify isometrically L with another vector space L^* of cross-section on X in such a way that the image $\overline{M}^* = \{0\}$. Namely, for each $f \in L$, let f^* be the cross-section $(f^*(x);\ x \in X)$ defined by

$$f^*(x) = f(x) + \overline{M}_x,$$

i.e. $f^*(x) \in E_x/\overline{M}_x$. Let L^* be the image of L under the linear mapping $f \to f^*$; L^* is a vector space of cross-section of the vector fibration $(X;(E_x/\overline{M}_x;\ x \in X))$. We endow L^* with the supremum norm:

$$||f^*|| = \sup_{x \in X} ||f^*(x)||$$

$$= \sup_{x \in X} \inf_{g \in M} ||f(x) - g(x)||.$$

By Theorem 1.30, $||f^*|| < +\infty$, and in fact

$$||f^*|| = \inf \{||f-g||;\ g \in M\}$$

Hence $f^* = 0$ if and only if $f \in \overline{M}$, and since the quotient norm in $L/\overline{M}$ is precisely $\inf \{||f-g||;\ g \in M\}$, the correspondence $f+\overline{M} \leftrightarrow f^*$ is an isometry between $L/\overline{M}$ and L^*.

It remains to prove that L^* is an essential vector space of upper semicontinuous cross-sections, which is a $C(X)$-module. Since

$$(gf)^*(x) = g(x)f(x) + \overline{M}_x =$$

$$= g(x)\left[f(x) + \overline{M}_x\right] = g(x)f^*(x)$$

if $g(x) \neq 0$, and $(gf)^*(x) = \overline{M}_x = 0.\ f^*(x)$ if $g(x) = 0$, we see immediately that L^* is a $C(X)$-module. To verify that L^* is essential, let v be any element of the fiber $E_x/\overline{M}_x$. Then $v = w+\overline{M}_x$, for some $w \in E_x$. Since L is essential, there is $f \in L$ with $f(x) = w$. But then $f^*(x) = f(x) + \overline{M}_x = w + \overline{M}_x = v$.

Finally, if $f \in L$, the quotient norm

$$||f^*(x)|| = \inf_{g \in M} ||f(x) - g(x)||$$

shows that $x \to ||f^*(x)||$ is an infimum of upper semicontinuous functions, and therefore it is upper semicontinuous too. This completes the proof of Theorem 1.38.

We come now to the case of A-modules $W \subset L$, where A is a self-adjoint subalgebra, not necessarily separating on X. However we restrain ourselves from the most general result and consider only the case $L = C(X;E)$, where X is a compact Hausdorff space and E is a normed space. Let Y be the quotient topological space of X modulo the equivalence relation $x \equiv y$ (mod.A), and let π be the quotient map of X onto Y. It follows that Y is a compact Hausdorff space and for each $a \in A$, there is a unique $b \in C(Y)$ such that $a = b \circ \pi$. If we put $B = \{b \in C(Y); a = b \circ \pi, a \in A\}$, then B is a self-adjoint subalgebra, which is separating over Y, containing the constants, if A contains the cosntants. By the Stone-Weierstrass Theorem (Corollary 1.9, §5), B is dense in $C(Y)$, if A contains the constants. On the other hand the mapping $a \to b$ is a linear isometry of A onto B. Hence B is closed, if A is closed. Therefore $B = C(Y)$, if A is a closed self-adjoint subalgebra containing the constants. For each $y \in Y$, let $E_y = C(\pi^{-1}(y); E)$. For each $f \in C(X;E)$, let $f^*(y) = f|\pi^{-1}(y)$, for all $y \in Y$. Then f^* is a cross-section of the vector fibration $(Y;(E_y;\ y \in Y))$. Let $L = \{f^*;\ f \in C(X;E)\}$. We claim that L is an upper semicontinuous space of cross-sections, when endowed with the norm

$$||f^*|| = \sup\{||f^*(y)||;\ y \in Y\}.$$

This is a consequence of Lemma 1.25, by taking there $g(x) = ||f(x)||$.

Since $\{\pi^{-1}(y);\ y \in Y\}$ is a closed partition of X, the mapping $f \to f^*$ is a linear isometry of $C(X;E)$ onto L. Notice also that L is essential, i.e. $L_y = \{f^*(y); f^* \in L\} = C(\pi^{-1}(y);E)$, when E is a Banach space. This follows from the Tietze Extension Theorem for Banach space-valued mappings on compact spaces (see Theorem 5.3, [52] or Theorem 3.4 below).

THEOREM 1.40 *Let $A \subset C(X)$ be a self-adjoint closed subalgebra, containing the constants and let $W \subset C(X;E)$ be an A-module where E is a Banach space and X is a compact Hausdorff space. Let $M = \{f^*;\ f \in W\}$. Then*

$$E(W^\perp \cap B') = \cup\{\delta_y(E(M_y^\perp \cap B_y'));\ y \in Y,\ \overline{M}_y \neq E_y\}.$$

PROOF To apply Theorem 1.38 we must show that both M and L are $C(Y)$-modules.

Indeed, let $f^* \in L$ and $b \in C(Y)$. Since $B = C(Y)$, there is $a \in A$ such that $b \circ \pi = a$. Hence

$$b(y)\ f^*(y) = a(x)\ f|_{\pi^{-1}(y)} =$$

$$= (af)|_{\pi^{-1}(y)} = (af)^*(y)$$

where $x \in \pi^{-1}(y)$ is any point. Therefore $bf^* \in L$. In particular,

if $f^* \in M$, then $f \in W$, so $af \in W$, and $(af)^* = bf^*$ belongs to M.

The result now follows from Theorem 1.38, identifying $C(X;E)$ and L by the isometry $f \leftrightarrow f^*$.

§ 11. REPRESENTATION OF VECTOR FIBRATIONS.

In this paragraph we prove two propositions which together add up to an interesting representation result. The importance of this result lies in the illuminating role it plays as regards semicontinuity assumptions such as (3) of § 9. Propositions 1.42 and 1.43 are due to L. Nachbin and appeared in [39] by his permission.

DEFINITION 1.41. *Let* $(X; (E_x ; x \in X))$ *be a vector fibration satisfying conditions*

(1) *and* (2) *of* § 9, i.e.,

(1) X *is compact;*

(2) *each* E_x *is a normed space, whose norm we denote by* $\|\cdot\|$.

Let $L \subset \pi(E_x ; x \in X)$ *be a vector space of cross-sections. A representation of* L *is a linear map*

$$r : L \to C(F;\mathbb{K})$$

where F *is a compact Hausdorff space provided with a continuous onto map* $\pi : F \to X$ *such that, for all* $f \in L$ *and* $x \in X$, *the equality*

$$\|f(x)\| = \sup\{|r(f)(y)| ; y \in \pi^{-1}(\{x\})\}$$

results.

PROPOSITION 1.42. *Under the conditions of Definition 1.41, if* L *admits a representation* r, *then*

$$f \mapsto \|f\| = \sup\{\|f(x)\| ; x \in X\}$$

is a norm on L, *and* r *is a linear isometry of* $(L, \|\cdot\|)$ *into*

$(C(F;\mathbb{K}), \|\cdot\|_F)$, *where* $\|\cdot\|_F$ *is the sup-norm. Moreover, for each* $f \in L$, *the function* $x \mapsto \|f(x)\|$ *is upper semicontinuous, i.e.,* L *satisfies condition* (3) *of* § 9.

PROOF: For each $f \in L$, write $\tilde{f} = r(f)$. By the definition of representation, the equality $\|f(x)\| = \sup\{|\tilde{f}(y)|;\ y \in \pi^{-1}(\{x\})\}$ is true for each $x \in X$. Since π is continuous and onto, $(\pi^{-1}(\{x\});\ x \in X)$ is a partition of F into non-empty, compact subsets; and since $\tilde{f} \in C(F;K)$ for each $f \in L$, it follows that

$$\sup\{\|f(x)\|;\ x \in X\} = \sup\{\sup\{|\tilde{f}(y)|;\ y \in \pi^{-1}(\{x\})\};x \in X\} =$$

$$= \sup\{|\tilde{f}(y)|;\ y \in F\} = \|\tilde{f}\|_F < \infty .$$

It is now clear that r is a linear isometry of $(L,\|\cdot\|)$ into $(C(F;\mathbb{K}),\|\cdot\|_F)$. Let f be a fixed element of L. Since $\tilde{f} = r(f) \in C(F;K)$ is related to $\|f\|: X = \pi(F) \to \mathbb{R}$ by the set of equations

$$\|f(x)\| = \sup\{|\tilde{f}(y)|;\ y \in \pi^{-1}(\{x\})\},\ x \in X,$$

the upper semicontinuity of $\|f\|$ follows from Lemma 1.25. The proof is now complete.

The following converse of Proposition 1.42 is true.

PROPOSITION 1.43. *Let* $(X;(E_x;x \in X))$ *and* L *be as above. Assume that* L *is essential and that* $x \mapsto \|f(x)\|$ *is upper semicontinuous, for each* $f \in L$. *Then there exists a representation of* L.

PROOF: The proof consists of the construction of a compact Hausdorff space F jointly with a continuous onto map $\pi : F \to X$ and a linear map $r : L \to C(F;\mathbb{K})$ such that

$$\|f(x)\| = \sup\{|(r(f))(y)|;\ y \in \pi^{-1}(\{x\})\}$$

for all $f \in L$ and all $x \in X$.

First, observe that the function $f \in L \to \|f\| = \sup\{\|f(x)\|\ ;\ x \in X\}$ is a norm on L. Let $\overline{B}(E_x')$, $x \in X$, and $\overline{B}(L')$ denote the closed unit balls of the continuous duals E_x', $x \in X$, and L', respectively, as topological subspaces of these continuous duals provided with the respective weak - star topologies $\sigma(E_x', E_x)$, $x \in X$, and $\sigma(L', L)$. $\overline{B}(E_x')$, $x \in X$, and $\overline{B}(L')$ are compact Hausdorff spaces by the Alaoglu - Bourbaki Theorem. F, purely as a set, is defined through the equation

$$F = \bigcup\{\{x\} \times \overline{B}(E_x')\ ;\ x \in X\},$$

F is a disjoint union, or sum, of the sets $\overline{B}(E_x')$, $x \in X$. This set is provided with the natural onto map $\pi : F \to X$ such that for each $(x, \phi) \in F$, $\pi(x, \phi) = x$. The bulk of the proof consists of providing F with a topology. Define $\omega : F \to \overline{B}(L')$ as follows. For each $(x, \phi) \in F$, that is, for each $x \in X$ and $\phi \in \overline{B}(E_x')$, the value of ω at (x, ϕ) is the functions $\omega_{x,\phi} : L \to \mathbb{K}$ such that $\omega_{x,\phi}(f) = \phi(f(x))$ for all $f \in L$. This definition of ω is justified by the following assertion:

(1) $\omega_{x,\phi} \in \overline{B}(L')$ for each $x \in X$ and each $\phi \in \overline{B}(C')$.

Indeed, $\omega_{x,\phi}$ is obviously linear and,

(2) for each $f \in L$, $|\omega_{x,\phi}(f)| = |\phi(f(x))| \leq \|f(x)\| \leq \|f\|$.

This computation also proves the necessity part of the next assertion:

(3) A linear functional Φ on L belongs to the image set $\omega(F)$ if, and only if, there exists $x \in X$ such that $|\Phi(f)| \leq \|f(x)\|$ for all $f \in L$. If this condition is satisfied, then $\Phi = \omega_{x,\phi}$ where $\phi \in \overline{B}(E_x')$ is such that for any given $z \in E_x'$, $\phi(z) = \Phi(f)$ for any $f \in L$ for which $f(x) = z$.

For let x and ϕ be as in (2) for a given Φ. The function ϕ is well - defined: indeed, if $f_1, f_2 \in L$ and $f_1(x) = f_2(x)$, then $|\Phi(f_1) - \Phi(f_2)| = |\Phi(f_1 - f_2)| \leq \|f_1(x) - f_2(x)\| = 0$;

on the other hand, given any $z \in E_x$ there is an $f \in L$ such that $f(x) = z$, because L is essential. ϕ is obviously a linear functional on E_x. Actually, $\phi \in \overline{B}(E_x')$; for, if $z \in E_x$ and $f \in L$ are such that $f(x) = z$, then $\phi(z) = \Phi(f(x))$, and therefore $|\phi(z)| = |\Phi(f(x))| \leq \|f(x)\| = \|z\|$ which proves that $\phi \in \overline{B}(E_x')$. For this ϕ, what is $\omega_{x,\phi}$? It is such that for all $z \in E_x$, or, since L is essential, for all $f(x)$, $f \in L$, it is true that $\omega_{x,\phi}(f) = \phi(f(x)) = \Phi(f)$. Hence $\Phi = \omega_{x,\phi} \in \omega(F)$ and the proof of (3) is complete.

To the maps $\pi : F \to X$ and $\omega : F \to \overline{B}(L')$ corresponds the map

$$\pi \times \omega \; F \to X \times \overline{B}(L').$$

(4) $\pi \times \omega$ is an injective map.

To see this let (x , ϕ) and (y , ψ) be elements of F such that $(x , \omega_{x,\phi}) = (y , \omega_{y,\psi})$. Then $x = y$; hence $\omega_{x,\psi} = \omega_{x,\psi}$, i. e., for all $f \in L$, $\omega_{x,\phi}(f) = \phi(f(x)) = \omega_{x,\psi}(f) = \psi(f(x))$ implying that $\phi = \psi$ (because $L(x) = E_x$).

Providing $X \times \overline{B}(L')$ with the product topology, it is then a compact Hausdorff space. The following assertion is true as a consequence (in particular) of the semicontinuity assumption.

(5) The image $(\pi \times \omega)(F)$ is a compact subset of the compact Hausdorff space $X \times \overline{B}(L')$.

It is enough to prove that $(\pi \times \omega)(F)$ is closed in $X \times \overline{B}(L')$. To this end let $(x_j , \omega_{x_j,\phi_j})$ be a net in $(\pi \times \omega)(F)$ convergent to $(y , \Phi) \in X \times \overline{B}(L')$. Since $X \times \overline{B}(L')$ has the product topology, it follows that $x_j \to y$ in X and that $\omega_{x_j,\phi_j} \to \Phi$ in $B(L')$. However, convergence in $\overline{B}(L')$ means weak convergence, therefore, for each $f \in L$,

$$\omega_{x_j,\phi_j}(f) = \phi_j(f(x_j)) \to \Phi(f)$$

results. Using (2), the inequality $|\omega_{x_j,\phi_j}(f)| \leq \| f(x_j) \|$

follows. Since $x_j \to y$ and $\|f\|$ is upper semicontinuous for each $f \in L$, it follows that

$$\lim_j |\omega_{x_j,\phi_j}(f)| = |\Phi(f)| \leq \limsup_j \|f(x_j)\| = \|f(y)\|$$

for all $f \in L$. (3) now implies that $\Phi \in \omega(F)$; actually, $\Phi = \omega_{y,\phi}$ where $\phi \in \overline{B}(E_y')$ with $\phi(z) = \Phi(f)$ for $z \in E_y$ and $f \in L$ with $f(y) = z$. Hence $(y, \Phi) = (y, \omega_{y,\phi}) \in \omega(F)$ proving that $\omega(F)$ is closed in $X \times \overline{B}(L')$.

By (4), $\pi \times \omega$ is an injective map so that the correspondence $(x, \phi) \in F \to (x, \omega_{x,\phi}) \in (\pi \times \omega)(F)$ is bijective. F is topologized by transporting to it the structure of a compact Hausdorff space of $(\pi \times \omega)(F)$ via this bijection. With each $f \in L$, associate $\tilde{f} : F \to \mathbb{K}$ defined by $\tilde{f}(x,\phi) = \omega_{x,\phi}(f) = \phi(f(x))$ for all $(x, \phi) \in F$. Assume that it has been proved that $\tilde{f} \in C(F;K)$ for all $f \in L$ and consider $r : L \to C(F;\mathbb{K})$ such that $r(f) = \tilde{f}$ for all $f \in L$. The map is obviously linear. As a consequence of the Hahn - Banach Theorem, the equality $\|f(x)\| = \sup\{|\phi(f(x))|;\ \phi \in \overline{B}(E_x')\}$ is obtained for any $x \in X$ and any $f \in L$; but, clearly,

$$\sup\{|\phi(f(x))|;\ \phi \in \overline{B}(E_x')\} = \sup\{|\tilde{f}(x,\phi)|;\ \phi \in \overline{B}(E_x')\} =$$

$$= \sup\{|(r(f))(y)|;\ y \in \pi^{-1}(\{x\})\},$$

since $\pi^{-1}(\{x\}) = \{x\} \times \overline{B}(E_x')$. Therefore r is a representation of L except that the assertion "$r(f) = \tilde{f} \in C(F;\mathbb{K})$ for all $f \in L$" still remains to be proved. The truth of the assertion follows from the following remarks.

By the definition of the topology of F, the map $\pi \times \omega$ is continuous. Fix $f \in L$. The evaluation map

$$\Phi \in \overline{B}(L') \to \Phi(f)$$

is continuous in the given topology of $\overline{B}(L')$; hence, $s_f : E \times \overline{B}(L') \to \mathbb{K}$ such that $s_f(x, \Phi) = \Phi(f)$ for all $(x, \Phi) \in E \times \overline{B}(L')$ is continuous. Finally, $\tilde{f} = r(f) = s_f \circ (\pi \times \omega)$ and the proof is complete.

§ 12. THE APPROXIMATION PROPERTY:

A locally convex space E has the *approximation property* if given $A \subset E$, a precompact subset, $p \in cs(E)$, and $\varepsilon > 0$, there is a continuous linear mapping T of finite rank from E into E, i.e. an element $T \in E' \otimes E$, such that $p(x - T(x)) < \varepsilon$ for all $x \in A$. When E is normed and the above T may be chosen with $\| T \| \leq 1$, we say that E has the *metric approximation property*.

In this section, X is a completely regular Hausdorff space and W is a vector subspace of $C(X)$ containing $C(X;[0,1])$.

THEOREM 1.44. *Let* $W \subset C(X)$ *be a vector subspace containing* $C(X;[0,1])$. *Then* W *with the compact - open topology has the approximation property. If* X *is a compact space, then* W *with the uniform topology has the metric approximation property.*

PROOF: Let $B \subset W$ be a precompact subset in the compact - open topology. Let $K \subset X$ be a compact subset and $p = |\cdot|$; and let $\varepsilon > 0$ be given. If X is a compact Hausdorff space take $K = X$ and $p = |\cdot|$. Since B is precompact, there is a finite set $F \subset B$ such that, given $f \in B$, there is $g \in F$ with $p(f(x) - g(x)) < \varepsilon / 3$ for all $x \in K$. For each $x \in K$, define

$$V_x = \{t \in X;\ p(g(t) - g(x)) < \varepsilon / 3,\ g \in F\}.$$

Since F is finite, V_x is an open neighborhood of x in X. By compactness, there is a finite set $\{x_1, x_2, \ldots, x_n\} \subset K$ such that $K \subset V_{x_1} \cup \ldots \cup V_{x_n}$.

Choose $\phi_1, \phi_2, \ldots, \phi_n \in C(X;\mathbb{R})$ such that

(1) $\phi_i(x) = 0$, if $x \notin V_{x_i}$ $(i = 1,2,\ldots,n)$;

(2) $0 \leq \phi_i(x) \leq 1$, for all $x \in X$ $(i = 1,2,\ldots,n)$;

(3) $\phi_1(x) + \phi_2(x) + \ldots + \phi_n(x) = 1$, for all $x \in K$,

(4) $\phi_1(x) + \phi_2(x) + \ldots + \phi_n(x) \leq 1$, for all $x \in X$.

Let us define $T : C(X) \longrightarrow C(X)$ by

$$(Tf)(x) = \sum_{i=1}^{n} \phi_i(x) f(x_i)$$

for all $f \in W$. Clearly, T is a finite rank linear operator. We claim that T is continuous, and if X is compact

Indeed, we have

$$\sup\{p((Tf)(x));\ x \in K\} \leq$$

$$\leq \sup\{\sum_{i=1}^{n} \phi_i(x) \cdot p(f(x_i));\ x \in K\} \leq$$

$$\leq \sup\{p(f(t));\ t \in K\} \cdot \sup\{\sum_{i=1}^{n} \phi_i(x);\ x \in K\}$$

$$= \sup\{p(f(t));\ t \in K\},$$

for all $f \in W$, by using (2) and (3) above.

Since $\phi_1, \phi_2, \ldots, \phi_n \in C(X;[0,1]) \subset W$, $Tf \in W$, for all $f \in W$.

Let now $f \in B$. There exists $g \in F$ such that $p(f(x) - g(x)) < \varepsilon / 3$ for all $x \in K$. Hence

$$p(f(x) - (Tf)(x)) = p(\sum_{i=1}^{n} \phi_i(x)(f(x) - f(x_i))) \leq$$

$$\leq \sum_{i=1}^{n} \phi_i(x)\{p(f(x) - g(x)) + p(g(x) - g(x_i)) + p(g(x_i) - f(x_i))\}$$

$$< \frac{1}{3} + \frac{1}{3} + \sum_{i=1}^{n} \phi_i(x) p(g(x) - g(x_i)).$$

for all $x \in K$. To evaluate the remaining sum on the right-hand side, let

$$I_x = \{1 \leq i \leq n;\ x \in V_{x_i}\}$$

and

$$J_x = \{1 \leq i \leq n;\ x \notin V_{x_i}\}.$$

For $i \in I_x$, $p(g(x) - g(x_i)) < \varepsilon / 3$ and for

$i \in J_x$, $\phi_i(x) = 0$, by (1) above. Hence

$$\sum_{i=1}^{n} \phi_i(x)p(g(x) - g(x_i)) < \frac{\varepsilon}{3}$$

for all $x \in K$. Thus $p(f(x) - (Tf)(x)) < \varepsilon$, for all $x \in K$, ending the proof.

COROLLARY 1.45. *If* X *is a compact Hausdorff space,* C(X) *has the metric approximation property.*

APPENDIX

NON-LOCALLY CONVEX SPACES

In this appendix we show how to extend some of the results of Chapter 1 to the case of vector-valued functions with values in a topological vector space E which is not locally convex. In this case, instead of the compact-open topology κ one considers in $C(X;E)$ a weaker topology, namely the topology κ_f of uniform convergence on compact subsets which have finite covering dimension.

A neighborhood basis of 0 for the topology κ_f is given by the sets of the form

$$N = \{f \in C(X;E);\ f(K) \subset U\}$$

when K ranges over the compact subsets of X which have finite covering dimension, and U ranges over a basis of neighborhoods of 0 in E.

As an example let us prove the following result due to A.H. Shuchat.

THEOREM 1 $C(X) \otimes E$ *is* κ_f*-dense in* $C(X;E)$.

PROOF Let $f \in C(X;E)$ be given. Let $K \subset X$ be a compact subset of X which has finite covering dimension, say n, and let $W \subset E$ be a neighborhood of 0 in E. Choose $U \subset E$ an open balanced neighborhood of 0 in E such that

$$U + \ldots + U \subset W$$

(where the sum is $(n+1)$- fold).

Each $x \in K$ lies in the open set $f^{-1}(f(x) + U)$. Consider the open covering of the compact set K consisting of $\{f^{-1}(f(x) + U);\ x \in K\}$. This open covering has a refinement of order at most $n + 1$ that is also an open covering of K. Let $V_1, \ldots, V_m$ be this refinement where $V_i = f^{-1}(f(x_i) + U)$, $i = 1, \ldots, m$, and $x_1, \ldots, x_m \in K$.

Let $g_1, \ldots, g_m$ be a continuous partition of the unity subordinate to this covering, i.e. each $g_i \in C(X)$, $0 \leq g_i \leq 1$, and

(1) $g_i(x) = 0$ if $x \notin V_i$ for all $i = 1,\ldots,m$;

(2) $\sum_{i=1}^{m} g_i(x) \leq 1$ for all $x \in X$;

(3) $\sum_{i=1}^{m} g_i(x) = 1$ for all $x \in K$.

(This is possible because we have assumed X to be completely regular).

Let $g = \sum_{i=1}^{m} g_i \otimes f(x_i)$. Then $g \in C(X) \otimes E$ and we claim that $(f-g)(K) \subset W$.

Indeed, if $x \in K$, then

$$f(x) - g(x) = \sum_{i=1}^{m} g_i(x) \ (f(x) - f(x_i))$$

Let $I_x = \{1 \leq i \leq m;\ x \in V_i\}$. By (1), we see that

$$f(x) - g(x) = \sum_{i=I_x} g_i(x)(f(x) - f(x_i)).$$

On the other hand, for each $i \in I_x$ we have $f(x)-f(x_i) \in U$, and since I_x has at most $n + 1$ elements, we see that

$$f(x) - g(x) \in U +\ldots+ U \subset W,$$

since U is balanced, and this ends the proof.

For further results on approximation in the non-locally convex case see the paper of A.H. Shuchat, "Approximation of vector-valued continuous functions", Proc. Amer. Math. Soc. 31 (1972), 97-103.

REFERENCES FOR CHAPTER 1.

Number in brackets refer to the Bibliography

ARENS [1]

ARENS and KELLEY [2]

BIERSTEDT [5]

BISHOP [8]

BRIEM, LAURSEN and PEDERSEN [9]

BROSOWSKI and DEUTSCH [10]

BUCK [12]

CHALICE [13]

CUNNINGHAM and ROY [15]

DUNFORD and SCHWARTZ [20]

GLICKSBERG [26], [27]

JEWETT [32]

MACHADO [37], [38]

MACHADO and PROLLA [39], [41]

NACHBIN [43]

PROLLA and MACHADO [52]

RUDIN [54], [55], [56]

SINGER [60]

STONE [62]

STRÖBELE [63]

CHAPTER 2

THE THEOREM OF DIEUDONNÉ

Let X_1 and X_2 be two completely regular Hausdorff spaces. We shall denote by $C(X_1) \otimes C(X_2)$ the vector subspace of $C(X)$, where X is the product space of X_1 and X_2, consisting of all finite sums of functions of the form

$$(x,y) \to f(x)\ g(y)$$

where $f \in C(X_1)$ and $g \in C(X_2)$.

THEOREM 2.1 *The vector subspace* $(C(X_1) \otimes C(X_2)) \otimes E$ *is dense in* $C(X;E)$.

PROOF Let $A = C(X_1) \otimes C(X_2)$ and let $W = A \otimes E$. It can be easily verified that A is a self-adjoint subalgebra of $C(X)$, containing the constants. Since both X_1 and X_2 are completely regular Hausdorff spaces, A is separating over X and $W(x) = E$, for each $x \in X$. By Corollary 1.13, W is dense in $C(X;E)$.

COROLLARY 2.2 $C(X_1) \otimes C(X_2)$ *is dense in* $C(X)$.

We can generalize Theorem 2.1 to any number of factors.

THEOREM 2.3 *Let* $X = \pi(X_i;\ i \in I)$ *be an arbitrary product of completely regular Hausdorff spaces. The vector subspace of* $C(X;E)$ *generated by all mappings of the form*

$$x = (x_i)_{i \in I} \to \left[\prod_{i \in J} g_i(x_i)\right] v_J$$

where $J \subset I$ *is a non-empty finite subset;* $g_i \in C(X_i)$, *for all* $i \in J$; *and* $v_J \in E$; *is dense in* $C(X;E)$.

PROOF Analogous to that of Theorem 2.1. Corollary 2.2 and Theorem 2.3 are due to Dieudonné (see [18]).

Let now E and F be two locally convex Hausdorff spaces (non-zero) and let τ be a topology on the tensor product $E \otimes F$ such that the canonical bilinear map of $E \times F$ into $(E \otimes F, \tau)$ is continuous. The upper bound of all such topologies is a locally convex topology, called the *projective* tensor product topology on $E \otimes F$; it is the finest locally convex topology on $E \otimes F$ for which the canonical bilinear map is continuous.

Given two completely regular Hausdorff spaces X and Y, let $f \in C(X;E)$ and $g \in C(Y;F)$. If $E \otimes F$ has a topology such that the canonical bilinear map is continuous, then

$$(x,y) \to f(x) \otimes g(y) = (f \otimes g)(x,y)$$

belongs to $C(X \times Y; E \otimes F)$.

THEOREM 2.4 *If $E \otimes F$ has a topology such that the canonical bilinear map is continuous, then $C(X;E) \otimes C(Y;F)$ is dense in $C(X \times Y; E \otimes F)$.*

PROOF $C(X;E) \otimes C(Y;F)$ denotes the vector subspace of $C(X \times Y; E \otimes F)$ consisting of all finite sums of mappings of the form $f \otimes g$, where $f \in C(X;E)$ and $g \in C(Y;F)$. Since

$$t(u \otimes v) = (tu) \otimes v = u \otimes tv$$

for all $t \in \mathbb{K}$, $u \in E$ and $v \in F$, one easily verifies that $(C(X) \otimes C(Y)) \otimes (E \otimes F)$ is contained in $C(X;E) \otimes C(Y;F)$, and the result follows from Theorem 2.1.

REMARK The above vector-valued version of the Dieudonné Theorem can be generalized to tensor products $E \otimes_A F$ of locally convex spaces E and F which are topological modules over some locally convex topological algebra A.

DEFINITION 2.5 *Let A be a locally convex topological algebra, and let M and N be two locally convex spaces which are topological modules over A. Then $M \otimes_A N$ is defined to be the quotient locally convex space $(M \otimes N)|D$, where $M \otimes N$ is the tensor product of the vector spaces M and N endowed with the projective tensor product topology, and D is the closed linear subspace of $M \otimes N$ spanned by elements of the form* $(ax \otimes y - x \otimes ay)$, $a \in A$,

$x \in M$, $y \in N$.

If $f : X \to M$ and $g : Y \to N$, then $f \otimes_A g$ denotes the map $(x,y) \to f(x) \otimes_A g(y)$ from $X \times Y$ into $M \otimes_A N$. If f and g are continuous, $f \otimes_A g$ is also continuous. Moreover, if $f \in C(X,M)$ and $g \in C(Y,N)$, then $f \otimes_A g$ belongs to the space $C(X \times Y, M \otimes_A N)$. We will denote by $C(X,M) \otimes_A C(Y,N)$ the vector subspace of $C(X \times Y, M \otimes_A N)$ consisting of all finite sums of mappings of the form $f \otimes_A g$, where $f \in C(X,M)$, $g \in C(Y,N)$.

THEOREM 2.6 $C(X,M) \otimes_A C(Y,N)$ *is dense in* $C(X \times Y, M \otimes_A N)$.

PROOF The subspace $C(X,M) \otimes_A C(Y,N)$ contains

$$[C(X) \otimes C(Y)] \otimes (M \otimes_A N).$$

By Corollary 2.2, $C(X) \otimes C(Y)$ is dense in $C(X \times Y)$. Hence, $C(X,M) \otimes_A C(Y,N)$ is dense in $C(X \times Y) \otimes (M \otimes_A N)$, which is dense in $C(X \times Y, M \otimes_A N)$.

DEFINITION 2.7 *Let* A *be a topological algebra. A topological module over* A, *also said a topological* A-*module, is a topological vector space* B *which is a (left or right) module over* A *in the usual algebraic sense, such that the bilinear map* $A \times B \to B$ *(whose value at* (a,b) *we will denote by* ab*) is continuous.*

If both A and B are locally convex spaces, the bilinear map $A \times B \to B$ is continuous if, and only if, given any continuous seminorm p on B, there exist a continuous seminorm q_1 on A, a continuous seminorm p_2 on B and a constant $k > 0$ such that $p(ab) \leq kq_1(a)\, p_2(b)$ for all $a \in A$, $b \in B$.

EXAMPLES 2.8 (a) Every topological vector space is a topological module over the scalar field.

(b) Every topological algebra with jointly continuous multiplication is a topological module over itself.

DEFINITION 2.9 *A net* $\{x_i\}$ *of elements of a topological algebra* A *such that* $x_i a \to a$ *for every* $a \in A$, *is called an approximate left unit. Similarly one defines approximate right and two-sided units.*

An approximate left (or right) unit is said to be

bounded if the net $\{x_i\}$ is bounded. If A is locally convex, $\{x_i\}$ is bounded if, and only if, $\sup \{q(x_i);\ i \in I\} < \infty$ for each continuous seminorm q on A. If moreover there exists a constant $K < +\infty$ such that $\sup \{q(x_i);\ i \in I\} \leq K$ for all continuous seminorms q which belong to a family of seminorms which determines the topology of A, then $\{x_i\}$ is said to be uniformly bounded.

REMARK 2.10 Let R be any ring and let M be a left R-module. If R has a unit element e, and if em = m for all $m \in M$, then M is said to be a unital module. This motivates the following.

DEFINITION 2.11 *Let* A *be a topological algebra with an approximate left unit* $\{a_i\}$ *and let* B *be a (left or right) topological module over* A. *If* $a_i b \to b$ *for all* $b \in B$, *then* B *is said to be an approximate left-unital module. Similarly one defines approximate right-unital modules. If* $\{a_i\}$ *is an approximate two-sided unit and* $a_i b \to b$ *for all* $b \in B$, *then* B *is said to be an approximate unital module.*

DEFINITION 2.12 *Let* A *be a topological algebra and let* B *be a (left or right) topological module over* A. *We say that* B *is an essential* A*-module if the vector space spanned by* $AB = \{ab;\ a \in A,\ b \in B\}$ *is dense in* B.

THEOREM 2.13 *Let* A *be a locally convex topological algebra with a bounded approximate left unit* $\{x_i\}$, *and let* B *be a locally convex space which is a (left or right) topological* A*-module. Then the following are equivalent:*

(a) B *is an essential* A*-module.*

(b) B *is an approximate left-unital module.*

PROOF It is evident that (b) implies (a). Let $b_0 \in B$. Given $\varepsilon > 0$ and p a continuous seminorm on B, there exist a continuous seminorm q on A, a continuous seminorm p' on B and a constant $k > 0$ such that $p(ab) \leq kq(a)\,p'(b)$ for all $a \in A$, $b \in B$. Since $\{x_i\}$ is bounded, there exists a constant $c > 0$ such that $q(x_i) \leq c$ for all $i \in I$. Since B is an essential A-module, there exists $a_1, \ldots, a_n \in A$ and $b_1, \ldots, b_n \in B$ such that

$$p(b_0 - \Sigma\, a_k b_k) < \varepsilon/3$$

and

$$p'(b_0 - \Sigma\, a_k b_k) < \varepsilon/3ck.$$

Choose now $j_0 \in I$ such that $i > j_0$ implies

$$q(a_k - x_i a_k) < \varepsilon/(3np'(b_k))$$

for all $1 \leq k \leq n$. Then for $i > j_0$ we have

$$p(b_0 - x_i b_0) \leq p(b_0 - \Sigma\, a_k b_k) + p(\Sigma a_k b_k - x_i(\Sigma a_k b_k))$$
$$+ p(x_i(\Sigma\, a_k b_k) - x_i b_0) <$$
$$< \varepsilon/3 + \Sigma p((a_k - x_i a_k)b_k) + q(x_i)p'(\Sigma a_k b_k - b_0)k$$
$$< \varepsilon/3 + \Sigma q(a_k - x_i a_k)p'(b_k) + c \cdot \varepsilon/3c$$
$$< \varepsilon/3 + n \cdot \varepsilon/3n + \varepsilon/3 = \varepsilon.$$

This ends the proof that (a) $\Longrightarrow$ (b).

REMARK 2.14 For many topological modules the properties (a) and (b) of Theorem 2.13 are equivalent to the stronger property

(c) $AB = \{ab;\ a \in A,\ b \in B\} = B$.

When (c) is valid we say that the A-module B has the factorization property. Clearly any unital module has the factorization property. COHEN proved that every Banach algebra with a bounded approximate unit has the factorization property, a result that was extended by HEWITT to bounded approximate left-unital Banach modules.

THEOREM 2.14 *Let* A *be a locally convex topological algebra and let* M *be a locally convex space which is an essential topological module over* A. *Then* $C(X,A) \otimes C(Y,M)$ *is dense in* $C(X \times Y, M)$.

PROOF Both A and M are $\mathbb{K}$-modules and $A \otimes_K M$ is the linear span of AM in M, which is dense in M, since M is essential.

REFERENCES FOR CHAPTER 2.

DIEUDONNÉ [18] , [70]

PROLLA [50]

STONE [62]

CHAPTER 3

EXTENSION THEOREMS FOR VECTOR-VALUED FUNCTIONS

In this chapter we shall prove several extension theorems for vector-valued functions defined on compact subsets of completely regular spaces. If $Y \subset X$ is a closed subset of a completely regular Hausdorff space, then Y is also a completely regular Hausdorff space with the relative topology. If E is a locally convex space, let T_Y be the restriction map $T_Y : C(X;E) \to C(Y;E)$. This map is obviously linear, and since every compact subset of Y is a compact subset of X, the map T_Y is continuous. Let $C_b(X;E)|Y$ be the image of $C_b(X;E)$ under T_Y.

THEOREM 3.1 *The vector subspace* $C_b(X;E)|Y$ *is dense in* $C(Y;E)$.

PROOF Let $A = C_b(X)|Y$. Obviously, A is a self-adjoint subalgebra of $C(Y)$, containing the constants. Since X is completely regular, A is separating over Y, and $W = C_b(X;E)|Y \subset C(Y;E)$ is such that $W(x) = E$, for each $x \in Y$. Now W is obviously an A-module. By Corollary 1.13, W is dense in $C(Y;E)$, as claimed.

The following result is fundamental in establishing extension theorems. Our proof follows closely De La Fuente [16], which is also the source of Definition 3.3 below.

LEMMA 3.2 *Let X be a compact Hausdorff space. For any non-empty closed subset $Y \subset X$, the linear mapping*

$$T_Y(f) = f|Y$$

from $C(X;E)$ *into* $C(Y;E)$ *is a topological homomorphism, for each locally convex space* E.

PROOF It is enough to prove that for some neighborhood base F of 0 in $C(X;E)$, $T_Y(U)$ is a relatively open subset of $C(X;E)|Y$, for all $U \in F$. Let us consider the neighborhood base F of 0 in $C(X;E)$ consisting of all subsets of the form

$$U = \{g \in C(X;E);\ p(g(x)) < \varepsilon,\ x \in X\}$$

where $p \in cs(E)$ and $\varepsilon > 0$. The subset W of $C(Y;E)$ defined by

$$W = \{h \in C(Y;E);\ p(h(x)) < \varepsilon,\ x \in Y\}$$

is then an open neighborhood of 0 in $C(Y;E)$. We claim that $T_X(U) = W \cap [C(X;E)|Y]$, whence $T_X(U)$ is relatively open in $C(X;E)|Y$. The inclusion $T_X(U) \subset W \cap [C(X;E)|Y]$ is obvious. Conversely, let $h \in W \cap [C(X;E)|Y]$. Let $g \in C(X;E)$ be such that $g(x) = h(x)$ for all $x \in Y$. Let $F = \{t \in X;\ p(g(t)) \geq \varepsilon\}$. Then $F \subset X$ is closed and disjoint from Y. If $F = \emptyset$, then $g \in U$, and therefore $h \in T_Y(U)$. If $F \neq \emptyset$, there exists $\phi \in C(X)$, $0 \leq \phi \leq 1$, $\phi(x) = 1$ for all $x \in Y$, and $\phi(t) = 0$ for all $t \in F$. Let $f = \phi g \in C(X;E)$. Then $f(x) = g(x) = h(x)$ for all $x \in Y$, i.e. $h = T_Y(f)$. We claim that $f \in U$. Let $x \in X$. If $x \in F$, then $f(x) = 0$, so $p(f(x)) = 0 < \varepsilon$. If $x \notin F$, then $p(f(x)) = p(\phi(x)g(x)) = \phi(x)p(g(x)) \leq p(g(x)) < \varepsilon$. Thus $g \in U$, and $h \in T_Y(U)$.

DEFINITION 3.3 *Let* C *be a class of locally convex Hausdorff spaces. We say that* C *has property* E_C *if:*

(1) *every element of* C *is complete;*

(2) *for every compact Hausdorff space* X, *and for every* $E \in C$, $C(X;E)$ *belongs to* C*;*

(3) *for every* $E \in C$, *and every closed vector subspace* $F \subset E$, *the quotient* E/F *belongs to* C.

EXAMPLES. (a) The class of all Banach spaces; (b) the class of all Fréchet spaces.

THEOREM 3.4 *(De La Fuente [16]). Let* C *be a class of locally convex Hausdorff spaces satisfying property* E_C. *Let* X *be a compact Hausdorff space and let* $Y \subset X$ *be a non-empty closed subset. Then* $C(X;E)|Y = C(Y;E)$ *for all* $E \in C$.

PROOF By Theorem 3.1, all we have to prove is that $C(X;E)|Y$ is closed in $C(Y;E)$. Let N be the kernel of T_Y. Since $E \in C$ and X is compact, $C(X;E)$ belongs to C. The linear mapping T_Y being continuous, N is a closed subspace of $C(X;E)$. By condition (3) of Definition 3.3, $C(X;E)/N$ belongs to C, and therefore by (1) of same Definition, $C(X;E)/N$ is complete. By Lemma 3.2, T_Y is a topological homomorphism. Hence $C(X;E)/N$ and $T_Y(C(X;E)) = C(X;E)|Y$ are topologically linearly isomorphic. Thus $C(X;E)|Y$ is complete too, and therefore closed in $C(Y;E)$. This ends the proof.

COROLLARY 3.5 *Let* C *be a class of locally convex Hausdorff spaces satisfying property* E_c. *Let* X *be a completely regular Hausdorff space and let* $Y \subset X$ *be a non-empty compact subset. Then* $C_b(X;E)|Y = C(Y;E)$, *for all* $E \in C$.

PROOF The space X is contained in its Stone-Čech compactification βX, and $Y \subset X$ being compact is closed in βX. By Theorem 3.4, $C(\beta X;E)|Y = C(Y;E)$. If $f \in C(Y;E)$, let $h \in C(\beta X;E)$ be such that $h(x) = f(x)$ for all $x \in Y$. Let now $g = h|X$. Then $g \in C_b(X;E)$ and $T_Y(g) = f$, i.e. $C_b(X;E)|Y = C(Y;E)$.

REMARK. The extension $g \in C_b(X;E)$ in Corollary 3.5 can be chosen so that $||g||_X = ||f||_Y$, when E is a Banach space. Indeed, let $r = ||f||_Y$. Define $\phi : E \to E$ by $\phi(t) = t$, if $||t|| \leq r$; and $\phi(t) = rt/||t||$, if $||t|| \geq r$. Let $g \in C_b(X;E)$ be an extension of f. Let $h = \phi \circ g$. Then $||h(x)|| = ||\phi(g(x))|| \leq r = ||f||_Y$ for all $x \in X$, i.e. $||h||_X \leq ||f||_Y$. On the other hand, for all $x \in Y$, $||f(x)|| = ||g(x)|| \leq r$; whence $\phi(g(x)) = g(x)$, i.e., $h(x) = f(x)$ for all $x \in Y$. Therefore h is an extension of f and $||f||_Y \leq ||h||_X$.

THEOREM 3.6 *Let* Y *be a closed non-empty subset of a normal Hausdorff space* X, *let* E *be a Banach space, and let* $f \in C(Y;E)$ *be a compact mapping. There exists a compact mapping* $g \in C(X;E)$ *such that* $g|Y = f$, *and* $g(X)$ *is contained in the closed convex hull of* $f(Y)$.

PROOF From the properties of the Stone-Čech compactification, it is known that βY identifies with the closure of Y in βX (see Stone [61]) and that a mapping $f \in C(Y;E)$ has an extension to βY if, and only if, $f(Y)$ is precompact. Hence we can extend f to βY. Call βf this extension $\beta f \in C(\beta Y;E)$. By Theorem 3.4 there exists $h \in C(\beta X;E)$ such that $h|\beta Y = \beta f$. Let $g' = h|X$. Then $g' \in C(X;E)$ and $g'|Y = f$. Let K be the closed convex hull of $f(Y)$. Since E is a Banach space, K is compact and, therefore, a retract of E (see Dugundji [19]). Let r be a retraction of E onto K. Then $g = r \circ g'$ belongs to $C(X;E)$ is a compact mapping such that $g|Y = f$, and $g(X) \subset K$.

REFERENCES FOR CHAPTER 3.

DE LA FUENTE [16]

DUGUNDJI [19]

PROLLA and MACHADO [52]

STONE [61], [62]

CHAPTER 4

POLYNOMIAL ALGEBRAS

§ 1 BASIC DEFINITIONS AND LEMMAS

The important notion of polynomial algebra was introduced by Pełczyński [47] in 1957, using multilinear transformations. An equivalent definition using polynomials was introduced by Wulbert (unpublished) (see Prenter [49]). Both definitions are more restrictive than the one given here, because they make use of all multilinear transformations and polynomials, respectively, whereas we only ask for invariance under composition with those of finite type, and do not assume that the polynomial algebra contains the constants. Moreover, the previous work on polynomial algebras was restricted to compact spaces X and Banach spaces E. A third equivalent definitions was introduced by Blatter [4], who considered $C_o(X;E)$, for locally compact X and Banach space E. We introduced our definition in [52], for completely regular spaces X and locally convex spaces E, where we proved the equivalence between the several possible definitions of polynomial algebras. Independently, De La Fuente [16] studied polynomial algebras too.

LEMMA 4.1 *Let* $W \subset C(X;E)$ *be a vector subspace (resp. closed vector subspace). The following properties are equivalent.*

(1) W *is invariant under composition with continuous linear maps of finite rank* $u \in E' \otimes E$.

(2) *For every* $\phi \in E'$, $v \in E$, $f \in W$, *then* $x \to \phi(f(x))v$ *belongs to* W.

(3) $A = \{\phi \circ f;\ \phi \in F',\ f \in W\}$ *is a vector subspace (resp. a closed vector subspace) of* C(X), *such that* $A \otimes E \subset W$.

PROOF (1) obviously implies (2).

Assume (2). The set A is clearly invariant under multiplication by any scalar $\lambda \in \mathbb{K}$. Let $\phi \circ f$ and $\psi \circ g$ be given in A. If $\phi = 0$, $\phi \circ f + \psi \circ g = \psi \circ g \in A$. If $\phi \neq 0$, choose $v \in E$, such that $\phi(v) = 1$. By (2), $x \to \phi(f(x))\,v$ and $x \to \psi(g(x))v$ belong to W. Let $h \in W$ be the function defined by $x \to [\phi(f(x)) + \psi(g(x))]v$. Then $\phi \circ h \in A$, and $\phi(h(x)) = [\phi(f(x)) + \psi(g(x))]\phi(v) = [\phi \circ f + \psi \circ g](x)$ for all $x \in E$. Therefore A is a vector subspace of C(X). Obviously, (2) implies $A \otimes E \subset W$, since W is a vector subspace. Suppose now that W is closed, and that $f \in \overline{A}$. Choose a pair $\phi \in E'$, $v \in E$, with $\phi(v) = 1$. Let $g = f \otimes v$. Given $K \subset X$ a compact subset, $\varepsilon > 0$, and p a continuous seminorm on E, there is $\psi \circ h \in A$ such that

$$|f(x) - \psi(h(x))| < \varepsilon/(1 + p(v))$$

for all $x \in K$. Hence $p(g(x) - \psi(h(x))v) < \varepsilon$, for all $x \in K$. Since $x \to \psi(h(x))v$ belongs to W, it follows that $g \in \overline{W} = W$. Since $f = \phi \circ g$, we see that $f \in A$, i.e. A is closed too.

Assume (3). Let $f \in W$ and $u \in E' \otimes E$. Suppose $u = \Sigma\, \phi_i \otimes v_i$, with $\phi_i \in E'$, $v_i \in E$, $i = 1,\dots,n$. Then $u \circ f = \Sigma\, h_i \otimes v_i$, where $h_i = \phi_i \circ f \in A$, for each $i = 1,\dots,n$. Hence, $u \circ f \in A \otimes E$. By (3), $A \otimes E \subset W$, and consequently $u \circ f \in W$, which proves (1).

COROLLARY 4.2 *Let* E *be a locally convex Hausdorff space with the approximation property. Let* $W \subset C(X;E)$ *be a vector subspace invariant under composition with elements of* $E' \otimes E$. *Let* $A = \{\phi \circ f;\ \phi \in E',\ f \in W\}$. *Then for every* $g \in C(X;E)$ *we have* $g \in \overline{W}$ *if, and only if,* $\phi \circ g \in A$, *for every* $\phi \in E'$.

PROOF The statement $g \in \overline{W} \Rightarrow \phi \circ g \in \overline{A}$, for every $\phi \in E'$, is always true, even when E does not have the approximation property.

Conversely, assume that E has the approximation property, and let $g \in C(X;E)$ be such that $\phi \circ g \in \overline{A}$, for all $\phi \in E'$. Let $K \subset X$, a compact subset, $\varepsilon > 0$ and $p \in cs(E)$ be given. Since g(K) is a compact subset of E, and E has the approximation property, there is $u \in E' \otimes E$ such that $p(t-u(t)) < \varepsilon$ for all

$t \in g(K)$. Let $u = \Sigma\, \phi_i \otimes v_i$, with $\phi_i \in E'$, $v_i \in E$, $i = 1,2,\ldots,n$. Then

$$p(g(x) - \sum_{i=1}^{n} \phi_i(g(x))v_i) < \varepsilon$$

for all $x \in K$. Hence g belongs to the closure of $\bar{A} \otimes E$. Since $\bar{A} \otimes E \subset \overline{A \otimes E}$, g belongs to $\overline{A \otimes E}$. By Lemma 4.1, $A \otimes E \subset W$. Consequently, $g \in \bar{W}$, QED.

REMARK. The above Corollary 4.2 is the most useful tool in proving vector-valued versions of Theorems known for scalar-valued functions, when the range space E has the approximation property. As an example, let us prove Mergelyan's Theorem for vector-valued functions in this case. We use the following notation. If $K \subset \mathbb{C}$ is a compact subset, $A(K;E)$ is the closed subspace of $C(K;E)$ of all functions holomorphic in the interior of K. Let $W = \mathcal{P}(\mathbb{C};E)|K$. It is easily verified that W is invariant under composition with elements $u \in E' \otimes E$. Moreover, if we assume that $E \neq \{0\}$, then

$$\mathcal{P}(\mathbb{C};\mathbb{C})\big|_K = \{\phi \circ f; \phi \in E', f \in W\}.$$

Suppose now that $\mathbb{C}\setminus K$ is connected, and let $g \in A(K;E)$. For each $\phi \in E'$, $\phi \circ g \in A(K;\mathbb{C})$. By Mergelyan's Theorem, $\phi \circ g$ belongs to closure of $\mathcal{P}(\mathbb{C};\mathbb{C})\big|_K$ in $C(K;\mathbb{C})$. By Corollary 4.2, g belongs then to the closure of W in $C(K;E)$. This proves Mergelyan's Theorem for E-valued functions, when E has the approximation property. The proof for general E due to Bierstedt [5] uses the fact that $A(K;\mathbb{C})$ has the approximation property, when $K \subset \mathbb{C}$ has a connected complement.

To give another illustration of Corollary 4.2, let E be a Banach space over $\mathbb{C}$, and let $U \subset E$ be a non-void open subset. If F is another Banach space over $\mathbb{C}$, $H(U;F)$ denotes the space of all holomorphic mappings from U to F. If $F = \mathbb{C}$, we write simply $H(U)$. We recall that $f: U \to F$ is holomorphic in U, if f possesses a Fréchet derivative at each point $\xi \in U$, i.e. at each $\xi \in U$, there exists $Df(\xi) \in L(E;F)$ such that

$$\lim_{\|h\|\to 0} \frac{\| f(\xi + h) - f(\xi) - Df(\xi)h \|}{\|h\|} = 0.$$

See Nachbin [44] for a reference on properties of holomorphic mappings. In particular f has a power series development and the Cauchy formulae are valid. For the first derivative, we get for any $\rho > 0$ and $x \in E$ such that $\xi + \lambda x \in U$ for every $\lambda \in \mathbb{C}$, $|\lambda| \leq \rho$:

$$Df(\xi)x = \frac{1}{2\pi i} \int_{|\lambda|=\rho} \frac{f(\xi+\lambda x)}{\lambda^2} d\lambda .$$

If $K \subset U$ is compact, we can find a $\rho > 0$ such that $\xi + \lambda x \in U$ for all $x \in K$, $\lambda \in \mathbb{C}$, $|\lambda| \leq \rho$. Hence

$$\sup_{x \in K} \|Df(\xi)x\| \leq \frac{1}{\rho} \cdot \sup_{t \in K_\xi} \|f(t)\|$$

where K_ξ is the compact subset of U,

$$K_\xi = \{\xi + \lambda x;\ x \in K,\ \lambda \in \mathbb{C},\ |\lambda| = \rho\}.$$

THEOREM 4.3 *(Aron and Schottenloher [3]). Let* F *be a complex Banach space. The following properties are equivalent:*

(a) F *has the approximation property.*

(b) *For every complex Banach space* E *and for every non-void open subset* $U \subset E$, *the space* $H(U) \otimes F$ *is dense in* $H(U;F)$ *in the compact-open topology.*

(c) $H(F) \otimes F$ *is dense in* $H(F;F)$ *in the compact-open topology.*

(d) *The identity map* $id_F : F \to F$ *belongs to the compact-open closure of* $H(F) \otimes F$ *in* $H(F;F)$.

PROOF. (a) $\Rightarrow$ (b). Let $W = H(U) \otimes F \subset C(U;F)$. Obviously, W is invariant under composition with elements of $F' \otimes F$. Let $A = \{\phi \circ f;\ \phi \in F',\ f \in W\}$. Then $A = H(U)$. Let now $g \in H(U;F)$. For every $\phi \in F'$, $\phi \circ g$ is holomorphic in U, whence $\phi \circ g \in A$. By Corollary 4.2, g belongs to the closure of $W = H(U) \otimes F$ in the compact-open topology.

(b) $\Rightarrow$ (c). Take $U = E = F$.

(c) $\Rightarrow$ (d). Obvious.

(d) $\Rightarrow$ (a). Let $\varepsilon > 0$ and $K \subset F$ compact be given. By Cauchy's formula, the seminorm

$$p(f) = \sup \{\|Df(0)x\|;\ x \in K\}$$

defined in $H(F;F)$, is continuous in the compact-open topology. By (d), there exists $g \in H(F) \otimes F$ such that $p(g-id_F) < \varepsilon$. Let $u = Dg(0)$. Then $u \in F' \otimes F$ and $||u(x) - x|| < \varepsilon$ for all $x \in K$, which proves (a).

DEFINITION 4.4 *Let* E *and* F *be two non-zero locally convex spaces. For each integer* $n \geq 1$, $\mathcal{P}_f^n(E;F)$ *denotes the vector subspace of* $C(E;F)$ *generated by the set of all maps of the form* $x \to [\phi(x)]^n v$, *where* $\phi \in E'$ *and* $v \in F$. *The elements of* $\mathcal{P}_f^n(E;F)$ *are called* n-*homogeneous continuous polynomials of finite type from* E *to* F.

The vector subspace of $C(E;F)$ generated by the union of all $\mathcal{P}_f^n(E;F)$, $n \geq 1$, and the constant maps, is denoted by $\mathcal{P}_f(E;F)$, and its elements are called *continuous polynomials of finite type from* E *to* F. When $F = \mathbb{K}$, we write simply $\mathcal{P}_f^n(E)$ and $\mathcal{P}_f(E)$, if no confusion may arise.

Notice also that, when $E = \mathbb{C}$, the space $\mathcal{P}_f(\mathbb{C};F)$ is the set of all complex polynomials with coefficients in F, that is the set of all functions of the form

$$z \to \sum_{n=0}^{m} a_n z^n$$

where $m \in \mathbb{N}$, $a_n \in F$, $n = 0,1,\ldots,m$. In this case we write $\mathcal{P}_f(\mathbb{C};F) = \mathcal{P}(\mathbb{C};F)$.

DEFINITION 4.5 *Let* E *and* F *be two non-zero locally convex spaces. For each integer* $n \geq 1$, $\mathcal{L}_f(^nE;F)$ *denotes the vector subspace of* $C(E^n;F)$ *generated by the set of all maps of the form* $(x_1,\ldots,x_n) \to \phi_1(x_1) \ldots \phi_n(x_1)v$, *where* $\phi_1,\ldots,\phi_n \in E'$ *and* $v \in F$. *The elements of* $\mathcal{L}_f(^nE;F)$ *are called* n-*linear continuous maps of finite type from* E^n *into* F.

LEMMA 4.6 ([52]) *Let* $W \subset C(X;E)$ *be a vector subspace. The following properties are equivalent.*

(1) *For each integer* $n \geq 1$, *given* $g_1,\ldots,g_n \in W$ *and* $T \in \mathcal{L}_f(^nE;E)$, *the function* $x \to T(g_1(x),\ldots,g_n(x))$ *belongs to* W.

(2) *For each integer* $n \geq 1$, *given* $g \in W$ *and*

$p \in \mathcal{P}_f^n(E;E)$, $p \circ g$ *belongs to* W.

(3) $A = \{\phi \circ f;\ \phi \in E',\ f \in W\}$ *is a subalgebra of* C(X) *such that* $A \otimes E \subset W$.

(4) W *is invariant under composition with continuous linear maps of finite rank, and there exists a continuous map* $P : E \times E \to E$ *and a vector* $v_0 \in E$ *such that*

(a) $P(v_0,v_0) \neq 0$;

(b) $P(av_0,bv_0) = ab\cdot P(v_0,v_0)$, *for all* $a,b \in \mathbb{K}$;

(c) *given* $f,g \in W$, *the function* $x \to P(f(x),g(x))$ *belongs to* W.

PROOF (1) obviously implies (2).

Assume (2). Taking $n = 1$, we see that W satisfies property (2) of Lemma 4.1. Hence A is a vector subspace of C(X) such that $A \otimes E \subset W$. Let $\phi \circ f$ and $\psi \circ g$ be given in A. Since

$$4(\phi \circ f)(\psi \circ g) = [\phi \circ f + \psi \circ g]^2 - [\phi \circ f - \psi \circ g]^2,$$

all that remains to prove is that $(\phi \circ f)^2$ belongs to A, for any $\phi \in E'$ and $f \in W$. If $\phi = 0$, there is nothing to prove. If $\phi \neq 0$, choose $v \in E$ such that $\phi(v) = 1$. Let $p(t) = [\phi(t)]^2 v$ for all $t \in E$. Then $p \in \mathcal{P}_f^2(E;E)$, and by (2), $p \circ f \in W$. Let $g = p \circ f$. Then $\phi(g(x)) = \phi(p(f(x))) = \phi([\phi(f(x))]^2 v) = [\phi(f(x))]^2 = (\phi \circ f)^2(x)$, for all $x \in X$, i.e. $(\phi \circ f)^2 \in A$, and A is an algebra.

Assume (3). By Lemma 4.1, W is invariant under composition with continuous linear maps of finite rank $u \in E' \otimes E$. Choose a pair $\phi \in E'$ and $v_0 \in E$ with $\phi(v_0) = 1$. Define $p:E\times E \to E$ by $P(s,t) = \phi(s)\phi(t)v_0$ for all $s,t \in E$. Then P is continuous and satisfies (a) and (b). Let $f,g \in W$, Since $\phi \circ f$ and $\phi \circ g$ belong to A, (3) implies that $(\phi \circ f)(\phi \circ g) \in A$, and $h = (\phi \circ f)(\phi \circ g) \otimes v_0 \in W$. However, $h(x) = P(f(x),g(x))$ for all $x \in X$, which ends the proof of (4).

Finally, assume (4). Let $n \geq 1$ be given. Let $g_1,\dots,g_n \in W$ and $T \in \mathcal{L}_f(^nE;E)$ also be given. If $n = 1$, $T \in E' \otimes E$, and $T \circ g_1 \in W$ because W is invariant under compo-

sition with elements of $E' \otimes E$. Suppose $n > 1$. Since W is a vector space, we may assume that T is of form $T(x_1,\ldots,x_n) = \phi_1(x_1) \ldots \phi_n(x_n)v$, where $\phi_i \in E'$, $i = 1,\ldots,n$, and $v \in E$. Assume (1) is true for $n - 1$. Then $(x_1,\ldots,x_{n-1}) \to \phi_1(x_1) \ldots \phi_{n-1}(x_{n-1})v_o$ belongs to $\mathcal{L}_f({}^{n-1}E;E)$, and therefore the mapping $x \to \phi_1(g_1(x)) \ldots \phi_{n-1}(g_{n-1}(x))\, v_o$ belongs to W. Call it h. Let $g = (\phi_n \circ g_n) \otimes v_o$. Then $g \in W$, and therefore $x \to P(h(x),g(x))$ belongs to W. Choose $\phi \in E'$ such that $\phi(P(v_o,v_o)) = 1$. Then $\phi \otimes v$ belongs to $E' \otimes E$ and $x \to \phi(P(h(x),g(x)))v$ belongs to W. However $\phi(P(h(x),g(x)))v =$
$\phi(P(\phi_1(g_1(x) \ldots \phi_{n-1}(g_{n-1}(x))v_o, \phi_n(g_n(x))v_o)) =$

$\phi_1(g_1(x)) \ldots \phi_{n-1}(g_{n-1}(x))\phi_n(g_n(x))\phi(P(v_o,v_o)) =$

$T(g_1(x),\ldots,g_{n-1}(x),g_n(x))$, for all $x \in X$, and the proof is complete.

DEFINITION 4.7 *A vector subspace $W \subset C(X;E)$ is called a polynomial algebra (of the $1^{\underline{st}}$ kind) if it has any of the equivalent properties (1)-(4) of Lemma 4.6. A vector subspace $W \subset C(X;E)$ is called a polynomial algebra of the $2^{\underline{nd}}$ kind if property (1) of Lemma 4.6 is true for all $T \in \mathcal{L}({}^nE;E)$ and all $n \geq 1$.*

A polynomial algebra W of the $1^{\underline{st}}$ kind (resp. of the $2^{\underline{nd}}$ kind) is called *self-adjoint* if the algebra $A = \{\phi \circ f;\ \phi \in E',\ f \in W\}$ is a self-adjoint subalgebra of $C(X)$; and it is called *every-where different from zero* if, for any $x \in X$, there is $g \in W$ such that $g(x) \neq 0$.

CONVENTION. By "polynomial algebra", we mean a polynomial algebra of the $1^{\underline{st}}$ kind.

LEMMA 4.8 *Let E and F be two non-zero locally convex Hausdorff spaces. Then*

(a) *The vector subspace generated by the union of all $\mathcal{P}^n_f(E;F)$, with $n \geq 1$, is a polynomial algebra.*

(b) *The vector subspace $\mathcal{P}_f(E;F)$ is a polynomial algebra.*

PROOF Let $W \subset C(E;F)$ be the vector subspace generated by the union of all $\mathcal{P}_f^n(E;F)$, with $n \geq 1$. Then $W \subset \mathcal{P}_f(E;F) = \mathcal{P}_f(E) \otimes F$. In fact,

$$W = \{p \in \mathcal{P}_f(E;F);\ p(0) = 0\}.$$

Therefore, Lemma 4.8 follows from Lemma 4.6 and the following

LEMMA 4.9 *For any non-zero locally convex space* E, *the vector space* $\mathcal{P}_f(E)$ *is an algebra.*

PROOF It is enough to prove that any product $\phi_1 \cdot \phi_2 \cdot \ldots \cdot \phi_m$ of m linear forms $\phi_i \in E'$ $(i = 1,2,\ldots,m)$ can be written as a linear combination of elements of $\mathcal{P}_f^m(E)$. By the "polarization formula" we have

$$(1) \quad x_1 \cdot \ldots \cdot x_m = \frac{1}{m!2^m} \Sigma\, \varepsilon_1, \ldots, \varepsilon_m (\varepsilon_1 x_1 + \ldots + \varepsilon_m x_m)^m,$$

where the summation is extended over all possible combinations of $\varepsilon_1 = \pm 1$, $\varepsilon_2 = \pm 1, \ldots, \varepsilon_m = \pm 1$, for all $x_1, x_2, \ldots x_m \in \mathbb{K}$.

Since $\varepsilon_1 \phi_1 + \ldots + \varepsilon_m \phi_m \in E'$,

$$x \to [\varepsilon_1 \phi_1(x) + \ldots + \varepsilon_m \phi_m(x)]^m$$

belongs to $\mathcal{P}_f^m(E)$, and therefore substituting $\phi_i(x)$ for x_i $(i = 1,2,\ldots,m)$ (1) yields

$$(2) \quad \phi_1(x) \ldots \phi_m(x) = \frac{1}{m!2^m} \Sigma\, \varepsilon_1 \ldots \varepsilon_m (\varepsilon_1 \phi_1(x) + \ldots + \varepsilon_m \phi_m(x))^m$$

for all $x \in E$.

As another example of a polynomial algebra $W \subset C(X;E)$ let us consider the following situation. Let E be a real finite-dimensional non-associative (i.e. not necessarily associative) linear algebra. This means that E is a finite-dimensional vector space over $\mathbb{R}$ in which a bilinear multiplication

$$(u,v) \in E \times E \to u\,v \in E$$

is defined. Since E is finite-dimensional there is only one locally convex and Hausdorff topology on E, and we shall always consider this topology for E. Notice that the multiplication being bilinear is then continuous.

By defining operations pointwise, $C(X;E)$ becomes a non-associative algebra over $\mathbb{R}$ too, as well as a *bimodule over* E : if $u \in E$ and $f \in C(X;E)$ the mappings $x \to u\, f(x)$ and $x \to f(x)u$ belong to $C(X;E)$. We shall call a vector subspace $W \subset C(X;E)$ a *submodule over* E if it is a bimodule over E, i.e. if it is invariant under right and left multiplication by elements of E.

LEMMA 4.10 *Let* E *be a real finite-dimensional central and simple non-associative linear algebra. Let* $W \subset C(X;E)$ *be a subalgebra over* $\mathbb{R}$ *which is a submodule over* E. *Then* W *is a polynomial algebra.*

Before proving Lemma 4.11 let us explain the terminology. All definitions are taken from Schafer [58]. An algebra E is called a *zero-algebra* if $uv = 0$ for all $u,v \in E$. The subspaces of E which are invariant relative to the right and left multiplications are called the *ideals* of E. The algebra E is called *simple* if E has no (two-sided) ideals $\neq 0$ and $\neq E$, and moreover E is not a zero-algebra. Let $\mathcal{M}(E)$ be the enveloping algebra of all right and left multiplications. $\mathcal{M}(E)$ is called the *multiplication algebra* of E. Clearly the ideals of E are the subspaces which are invariant relative to the multiplication algebra $\mathcal{M}(E)$. It follows that a non-zero algebra is simple if and only if $\mathcal{M}(E)$ is an irreducible algebra of linear transformations. We define the *centroid* of E to be the centralizer of $\mathcal{M}(E)$ in the algebra $\mathcal{L}(E)$ of all linear transformations on E. It follows that $T \in \mathcal{L}(E)$ belongs to the centroid of E if and only if

$$T(uv) = T(u).v = u.T(v)$$

for all $u,v \in E$. Clearly, all T of the form $T = \lambda \cdot id_E$, for $\lambda \in \mathbb{R}$, belong to the centroid. We say that E is *central* if its centroid coincides with $\mathbb{R}.id_E$. We have then the following fundamental result

LEMMA 4.11 *Let* E *be a real finite-dimensional central and simple non-associative algebra. Then* $\mathcal{M}(E) = \mathcal{L}(E)$.

PROOF Let Γ be the centroid of E. Then Γ is isomorphic to $\mathbb{R}$. The result follows from Theorem 4, Chapter X, Jacobson [31].

PROOF OF LEMMA 4.10 By Lemma 4.11, $\mathcal{M}(E) = \mathcal{L}(E)$. Therefore, any $W \subset C(X;E)$ which is a submodule over the algebra E is invariant under composition with any linear transformation $T \in \mathcal{L}(E)$.

Since E is not a zero-algebra, choose a pair $u,v \in E$ such that $uv \neq 0$. Let $\phi \in E'$ be a linear functional such that $\phi(uv) = 1$. Define $A = \{\psi(g);\ \psi \in E',\ g \in W\}$. By Lemma 4.1, A is a vector subspace of $C(X;\mathbb{R})$ such that $A \otimes E \subset W$. It remains to prove that A is a subalgebra. Let $\psi(g)$ and $\eta(h)$ be in A. Then $x \to \psi(g(x))u$ and $x \to \eta(h(x))v$ belong to W, since $A \otimes E \subset W$. By hypothesis, W is a subalgebra of $C(X;E)$ under pointwise operations. Thus the mapping $x \to [\psi(g(x))u][\eta(h(x))v] = \psi(g(x))\eta(h(x))uv$ belongs to W. Call it f. Then $\phi(f) \in A$. Clearly, $\phi(f(x)) = \psi(g(x))\eta(h(x))$ for all $x \in X$, since $\phi(uv) = 1$. Thus W is a polynomial algebra.

REMARK The above proof of Lemma 4.10 can be applied to any non-zero algebra such that $\mathcal{M}(E) = \mathcal{L}(E)$. In his Thesis [16], De La Fuente proved that $\mathcal{M}(E) = \mathcal{L}(E)$ for the following classes of algebras:

(1) E a Clifford algebra of a real vector space of even dimension;

(2) E a Cayley-Dickson algebra D_n, with $n \geq 2$.

In his monograph [4], Blatter assumes E to have a non-zero square, i.e. assumes the existence of an element $v \in E$ such that $v^2 \neq 0$. Thus his result cannot be applied to Lie algebras. A non-associative algebra E is said to be a *Lie algebra* if its multiplication satisfies the two conditions

(i) $v^2 = 0$

(ii) $(uv)w + (vw)u + (wu)v = 0$ for all $u,v,w \in E$.

From (i) and (ii) (known as the *Jacobi identity*) it follows that for all $u,v \in E$.

(iii) $uv = -vu$

Conversely, if the field over which E is a vector space is of characteristic $\neq 2$, then (iii) implies (i).

§ 2 STONE-WEIERSTRASS SUBSPACES

Motivated by the Stone-Weierstrass Theorem (Corollary 1.9, §5, Chapter 1) we state the following.

DEFINITION 4.12 *Let* $W \subset C(X;E)$ *be a vector subspace. The Stone-Weierstrass hull of* W *in* $C(X;E)$, *denoted by* $\Delta(W)$, *is the set of all functions* $f \in C(X;E)$ *such that*

(1) *for any* $x \in X$ *such that* $f(x) \neq 0$, *there is* $g \in W$ *such that* $g(x) \neq 0$;

(2) *for any* $x,y \in X$ *such that* $f(x) \neq f(y)$, *there is* $g \in W$ *such that* $g(x) \neq g(y)$.

Obviously, $\Delta(W) \subset C(X;E)$ is a vector subspace, containing W. Moreover, if E is a Hausdorff space, $\overline{W} \subset \Delta(W)$.

DEFINITION 4.13 *Let* $W \subset C(X;E)$ *be a vector subspace. We say that* W *is a Stone-Weierstrass subspace if* $\Delta(W) \subset \overline{W}$.

Before proceeding, let us show that $\Delta(W)$ is in fact a self-adjoint closed polynomial algebra containing W. To do this let us introduce the following function $\delta_W: R \to \{0,1\}$ (see Blatter [4]) :

a) $R \subset X \times X$ is the set of all pairs (x,y) such that $x \equiv y \pmod{W}$.

b) $\delta_W(x,y) = 0$, if $f(x) = 0$ for all $f \in W$.

c) $\delta_W(x,y) = 1$, if $f(x) \neq 0$ for some $f \in W$.

It is clear that the following property holds: $(x,y) \in R \Rightarrow f(x) = \delta_W(x,y)f(y)$ for all $f \in W$. Let $\Delta_1(W)$ be the set of all $g \in C(X;E)$ such that $(x,y) \in R \Rightarrow g(x) = \delta_W(x,y)g(y)$. Clearly, $W \subset \Delta_1(W)$.

PROPOSITION 4.14 *For every vector subspace* $W \subset C(X;E)$, $\Delta(W) = \Delta_1(W)$.

PROOF Let $f \in \Delta_1(W)$. Let $x \in X$ be such that $f(x) \neq 0$. If $g(x) = 0$ for all $g \in W$ then $\delta_W(x,x) = 0$, and $f(x) = \delta_W(x,x)f(x) = 0$, a contradiciton. This proves (1) of Definition 4.12. Let

$x,y \in X$ be such that $f(x) \neq f(y)$. Assume $g(x) = g(y)$ for all $g \in W$. Then $(x,y) \in R$. Since $f(x) \neq f(y)$, we may assume $f(x) \neq 0$. By (1) just proved, there is $g \in W$ with $g(x) \neq 0$. Hence $\delta_W(x,y) = 1$. Therefore $f(x) = \delta_W(x,y)f(y) = f(y)$, a contradiction. This proves (2) of Definition 4.12, and so $\Delta_1(W) \subset \Delta(W)$.

Conversely, assume $f \in \Delta(W)$. Let $(x,y) \in R$. Suppose that $\delta_W(x,y) = 0$. Then $g(x) = g(y) = 0$ for all $g \in W$. Since $f \in \Delta(W)$, $f(x) = f(y) = 0$. Suppose now that $\delta_W(x,y) = 1$. If $f(x) \neq f(y)$, there would exist $g \in W$ with $g(x) \neq g(y)$, which contradicts $(x,y) \in R$. Hence $f(x) = f(y)$. In both cases, $f(x) = \delta_W(x,y)\ f(y)$, and therefore $f \in \Delta_1(W)$.

PROPOSITION 4.15 *For every vector subspace* $W \subset C(X;E)$, $\Delta(W)$ *is a closed self-adjoint polynomial algebra containing* W.

PROOF Since $\delta_W(x,y) \in \{0,1\}$ for all $(x,y) \in R$, $\Delta_1(W)$ is obviously a polynomial algebra, containing W, such that $\{\phi \circ g;\ \phi \in E',\ g \in \Delta_1(W)\}$ is self-adjoint. Let $g \in \overline{\Delta_1(W)}$, and let $\{f_\alpha\}$ be a net, $f_\alpha \to g$, $f_\alpha \in \Delta_1(W)$. Let $(x,y) \in R$. Since $K = \{x,y\}$ is compact, and for every α, $f_\alpha(x) = \delta_W(x,y)f_\alpha(y)$, we see that $g \in \Delta_1(W)$. It remains to notice $\Delta_1(W) = \Delta(W)$ by the preceding Proposition 4.14.

LEMMA 4.16 *Let* $W \subset C(X;E)$ *be a vector subspace which is invariant under composition with any element* $u \in E' \otimes E$, *and let* $A = \{\phi \circ f;\ \phi \in E',\ f \in W\}$. *Suppose that* E *is a Hausdorff space. Then*

$$\Delta(W) = L_A(A \otimes E) = L_A(W).$$

PROOF Let $f \in \Delta(W)$. Let $Y \subset X$ be an equivalence class (mod.A). Let $x,y \in Y$. If $f(x) \neq f(y)$, there is $g \in W$ such that $g(x) \neq g(y)$. By the Hahn-Banach Theorem, there is $\phi \in E'$ such that $\phi(g(x)) \neq \phi(g(y))$. Since $\phi \circ g \in A$, this is impossible. Hence f is constant over Y. Let $v \in E$ be this constant value. If $v = 0$, then f agrees with $0 \in A \otimes E$ over Y. If $v \neq 0$, choose $g \in W$ such that $g(x) \neq 0$, for some $x \in Y$. Notice that g is constant over Y, since A and W define the same equivalence relation over X.

Let $u \in E$, $u \neq 0$, be this constant value. Choose $\phi \in E'$ with $\phi(u) = 1$. Then $h = (\phi \circ g) \otimes v$ belongs to $A \otimes E$ and agrees with f over Y. Hence $f \in L_A(A \otimes E)$.

By Lemma 4.1, §1, $A \otimes E \subset W$. Therefore $L_A(A \otimes E) \subset L_A(W)$.

Finally, let $f \in L_A(W)$. Let $x \in X$ be such that $f(x) \neq 0$, Suppose $g(x) = 0$ for all $g \in W$. Let $Y \subset X$ be the equivalence class (mod. A) that contains x. For every $\varepsilon > 0$ and $p \in cs(E)$ there is $g \in W$ such that $p(f(x) - g(x)) < \varepsilon$. Hence $p(f(x)) < \varepsilon$. Since E is Hausdorff, $f(x) = 0$. This contradiction shows that f satisfies (1) of Definition 4.12. Similarly, one proves that f satisfies condition (2) of Definition 4.12. So $f \in \Delta(W)$. This completes the proof of Lemma 4.16.

THEOREM 4.17 *(Stone-Weierstrass Theorem for polynomial algebras). Suppose* E *is a Hausdorff space. Every self-adjoint polynomial algebra* $W \subset C(X;E)$ *is a Stone-Weierstrass subspace.*

PROOF By Lemma 4.16, $\Delta(W) = L_A(W) = L_A(A \otimes E)$. By Theorem 1.8, § 5, Chapter 1, applied to the A-module $A \otimes E$, we have $L_A(A \otimes E) = \overline{A \otimes E}$. Since W is a polynomial algebra, $A \otimes E \subset W$. Hence $\overline{A \otimes E} \subset \overline{W}$. Putting all this together, $\Delta(W) \subset \overline{W}$, i.e. W is a Stone-Weierstrass subspace.

COROLLARY 4.18 *Suppose* E *is a Hausdorff space. Let* $W \subset C(X;E)$ *be a self-adjoint polynomial algebra. Then* W *is dense if and only if* W *is separating and every-where different from zero.*

PROOF Just notice that if W is separating and everywhere different from zero, then $\Delta(W) = C(X;E)$. Conversely, every dense subset of $C(X;E)$ is separating and everywhere different from zero, since E is Hausdorff.

COROLLARY 4.19 (Nachbin, Machado, Prolla [46]) *(Infinite dimensional Weierstrass polynomial approximation Theorem). Let* E *and* F *be two non-zero real locally convex Hausdorff spaces. Then* $\mathcal{P}_f(E;F)$ *is dense in* $C(E;F)$. *Moreover the vector subspace gene-*

rated by all $\mathcal{P}_f^n(E;F)$, *with* $n \geq 1$, *is dense in the polynomial algebra* $\{f \in C(E;F);\ f(0) = 0\}$.

PROOF By Lemma 4.8, § 1, $\mathcal{P}_f(E;F)$ is a polynomial algebra. Since E and F are real, $A = \{\phi \circ g;\ \phi \in F',\ g \in \mathcal{P}_f(E;F)\}$ is a subalgebra of $C(E;\mathbb{R})$. Since $\mathcal{P}_f(E;F)$ contains the constants and is separating over E (because both E and F are non-zero), Corollary 4.18 above shows that $\mathcal{P}_f(E;F)$ is dense in $C(E;F)$.

Let W be the vector subspace of $C(E;F)$ generated by the union of all $\mathcal{P}_f^n(E;F)$ with $n \geq 1$. By Lemma 4.8, § 1, W is a polynomial algebra. Let $A = \{\phi \circ g;\ \phi \in F',\ g \in W\}$. Since both E and F are real, $A \subset C(E;\mathbb{R})$. Let $f \in C(E;F)$ be such that $f(0) = 0$. Let $x \in E$ be such that $f(x) \neq 0$. Hence $x \neq 0$. Let $\phi \in E'$ with $\phi(x) = 1$ and let $v \in F$ with $v \neq 0$. Then $g = \phi \otimes v$ belongs to W and $g(x) = v \neq 0$. Let $x, y \in E$ be such that $f(x) \neq f(y)$. Hence $x \neq y$. Choose $\phi \in E'$ with $\phi(x) \neq \phi(y)$ and $v \in F$ with $v \neq 0$. Then $g = \phi \otimes v$ belongs to W and $g(x) \neq g(y)$. This shows that $f \in \Delta(W)$. By Theorem 4.17, $f \in \overline{W}$, as desired.

REMARK Corollary 4.19 has an analogue for *complex* spaces, if we redefine $\mathcal{P}_f^n(E) \otimes F$ as the vector subspace generated by the set of all maps of the form $x \to [\phi(x)]^n v$, where $v \in F$ and $\phi : E \to \mathbb{C}$ is either a linear or an antilinear continuous form. Let $\mathcal{P}_f^*(E;F)$ be the vector subspace generated by all $\mathcal{P}_f^n(E) \otimes F$, $n \geq 1$, defined as above. Then $A = \{\phi \circ g;\ \phi \in F',\ g \in \mathcal{P}_f^*(E) \otimes F\} = \mathcal{P}_f^*(E)$ is a self-adjoint subalgebra of $C(E;\mathbb{C})$.

COROLLARY 4.20 *Suppose E is a Hausdorff space. For every vector subspace* $W \subset C(X;E)$, $\Delta(W)$ *is the smallest closed self-adjoint polynomial algebra containing* W.

PROOF By Proposition 4.15, $\Delta(W)$ is a closed self-adjoint polynomial algebra containing W. Let $V \subset C(X;E)$ be a closed self-adjoint polynomial algebra containing W. Hence $\Delta(W) \subset \Delta(V)$. By Theorem 4.17, $\Delta(V) \subset \overline{V} = V$. Therefore $\Delta(W) \subset V$, as desired.

COROLLARY 4.21 *(Blatter[4]) Let* E *be a finite-dimensional central and simple non-associative real algebra. Every real sub-*

algebra $W \subset C(X;E)$ *which is a submodule over* E *is a Stone-Weierstrass subspace.*

PROOF By Lemma 4.10, § 1, W is a polynomial algebra. Hence we may apply Theorem 4.17.

COROLLARY 4.22 *(De La Fuente [16]) Let* E *be a Clifford algebra of a real vector space of even dimension or a Cayley-Dickson algebra* D_n, *with* $n \geq 2$. *Every real subalgebra* $W \subset C(X;E)$ *which is a submodule over* E *is a Stone-Weierstrass subspace.*

PROOF As noticed in the final Remark of § 1, we can apply Lemma 4.10, § 1. Therefore W is a polynomial algebra, and by Theorem 4.17 above, W is a Stone-Weierstrass subspace.

THEOREM 4.23 *Suppose* E *is a non-zero Hausdorff space. Let* $W \subset C(X;E)$ *be a vector subspace which is invariant under composition with elements of* $E' \otimes E$, *and let* $A=\{\phi o f;\ \phi \in E', f \in W\}$. *The following conditions are equivalent:*

(1) W *is localizable under* A *in* $C(X;E)$.

(2) W *is a Stone-Weierstrass subspace.*

(3) A *is a Stone-Weierstrass subspace.*

PROOF By Lemma 4.16, $\Delta(W) = L_A(W)$. Hence (1) and (2) are equivalent.

Assume (2), and let $f \in C(X;\mathbb{K})$ be an element of $\Delta(A)$. Let $K \subset X$ compact and $\varepsilon > 0$ be given. Choose $\phi \in E'$, with $\phi \neq 0$, and choose $v \in E$ with $\phi(v) = 1$. Let $g = f \otimes v$. Obviously $g \in \Delta(W)$. By hypothesis $g \in \overline{W}$. Let $p \in cs(E)$ be such that $|\phi(t)| \leq p(t)$ for all $t \in E$. Let $h \in W$ be chosen so that $p(g(x) - h(x)) < \varepsilon$ for all $x \in K$. Hence $|f(x) - (\phi \circ h)(x)| < \varepsilon$ for all $x \in K$. But $\phi \circ h \in A$, so $f \in \overline{A}$ and A is a Stone-Weierstrass subspace.

Finally, assume (3). Since $\overline{A} = \Delta(A)$, it follows from Proposition 4.15, that $B = \overline{A}$ is a closed self-adjoint subalgebra of $C(X;\mathbb{K})$. By Theorem 4.17 applied to the polynomial algebra $B \otimes E$, we have $L_B(B \otimes E) = \overline{B \otimes E}$. Hence $L_A(W) = L_A(A \otimes E) \subset L_A(B \otimes E) = L_B(B \otimes E) = \overline{B \otimes E} \subset \overline{A \otimes E}$, by Lemma 4.16 and the fact that $\bar{A} \otimes E \subset \overline{A \otimes E}$. By Lemma 4.1, § 1, $A \otimes E$ is contained in W; hence $L_A(W) \subset \bar{W}$, which proves (1).

We come now to Bishop's Theorem for polynomial algebras of the $2^{\underline{nd}}$ kind.

THEOREM 4.24 *Let* X *be a compact Hausdorff space and let* E *be a semi-normed space. Let* $W \subset C(X;E)$ *be a polynomial algebra of the* $2^{\underline{nd}}$ *kind and let* $A = \{\phi \circ f;\ \phi \in E',\ f \in W\}$. *For every* $f \in C(X;E)$, f *belongs to the closure of* W, *if and only if*, $f|S$ *belongs to the closure of* $W|S$ *in* $C(S;E)$, *for each maximal* A-*antisymmetric subset* $S \subset X$.

PROOF By Lemma 4.6, § 1, $A \subset C(X;\mathbb{C})$ is a subalgebra. For every $f, g \in W$ and $\phi \in E'$, the function $x \to \phi(f(x))g(x)$ belongs to W, since $(u,v) \to \phi(u)v$ belongs to $\mathcal{L}(^2E;E)$. Hence, W is an A-module. It remains to apply Theorem 1.27 (§ 8, Chapter 1).

§ 3 C(X)-MODULES

In this section we shall suppose throughout that E is a locally convex Hausdorff space.

Let $S \subset C(X;E)$ be an arbitrary subset and let us define

$$Z(S) = \{x \in X;\ g(x) = 0 \text{ for all } g \in S\}.$$

Obviously, Z(S) is a closed subset of X. On the other hand, if $Z \subset X$ is any closed subset let

$$I(Z) = \{f \in C(X;E); f(x) = 0 \text{ for all } x \in Z\}.$$

It is easy to check that, for any subset $S \subset C(X;E)$, $W = I(Z(S))$ is a closed polynomial algebra, containing S, which is a C(X)-module. Moreover, $A = \{\phi \circ f;\ \phi \in E',\ f \in W\}$ is self-adjoint. Indeed, let $g \in A$, say $g = \phi \circ f$, with $\phi \in E'$, $f \in W$. Choose a pair $\psi \in E'$ and $v \in E$ with $\psi(v) = 1$. Let $h = \bar{g} \otimes v$. Let $x \in Z(S)$. Then $g(x) = 0$, and $h(x) = \overline{g(x)}v = 0$, i.e. $h \in W$. Since $\bar{g} = \psi \circ h$, it follows that $\bar{g} \in A$, i.e. A is self-adjoint.

Let V be a closed polynomial algebra, containing S,

and such that

(1) V is a C(X)-module;

(2) $\{\phi \circ f;\ \phi \in E',\ f \in V\}$ is self-adjoint.

We claim that $W = I(Z(S)) \subset V$. Indeed, let $f \in W$. Let $x \in X$ be such that $f(x) \neq 0$. Then $x \notin Z(S)$, i.e. there exist $g \in S \subset V$ such that $g(x) \neq 0$. Let $x,y \in X$ be such that $f(x) \neq f(y)$. Then x and y do not belong simultaneously to $Z(S)$. Suppose $x \notin Z(S)$. Since X is a completely regular Hausdorff space, there is $h \in C(X)$ such that $h(x) = 1$, $h(y) = 0$. Let $g \in S \subset V$ be such that $g(x) \neq 0$. Then $hg \in V$, since V is a C(X)-module, and $h(x)g(x) = g(x) \neq 0 = h(y)g(y)$. By Theorem 1, § 2, $f \in \overline{V} = V$.

If $S = C(X;E)$, then $Z(S) = \emptyset$. Conversely, if W is a closed polynomial algebra satisfying (1) and (2) and such that $Z(W) = \emptyset$, then $W = I(Z(W)) = I(\emptyset) = C(X;E)$. This proves the following.

THEOREM 4.25 *Let* $S \subset C(X;E)$ *be an arbitrary subset, and* $W = I(Z(S))$. *Then* W *is the smallest closed polynomial algebra containing* S *and such that*

(1) W *is a* C(X)-*module;*

(2) $\{\phi \circ f;\ \phi \in E',\ f \in W\}$ *is self-adjoint.*

Moreover, W *is a closed polynomial algebra satisfying* (1) *and* (2) *if, and only if,* $W = I(Z(W))$. *A closed polynomial algebra* W *satisfying* (1) *and* (2) *is characterized by the associated closed set* $Z(W)$. *In particular,* $W = C(X;E)$ *if, and only if,* $Z(W) = \emptyset$.

COROLLARY 4.26 *The maximal proper closed self-adjoint polynomial algebras which are* C(X)-*modules are of the form*

$$W = \{f \in C(X;E);\ f(x) = 0 \text{ for some } x \in X\}.$$

COROLLARY 4.27 *Every proper closed self-adjoint polynomial algebra* W, *which is a* C(X)-*module, is contained in some maximal proper closed self-adjoint polynomial algebra which is a* C(X)-*module; in fact,* W *is the intersection of all the maximal proper closed self-adjoint polynomial algebras which are* C(X)-*modules and contain it.*

§ 4 APPROXIMATION OF COMPACT OPERATORS

If E and F are Banach spaces, let $L_c(E;F)$ be the uniform closure in the space of bounded *linear* operators from E to F of the set $E' \otimes F$ of continuous linear operators of finite rank from E to F. The space $L_c(E;F)$ is the space of *compact* linear operators from E to F if either E' or F has the approximation property. In this case if $u : E \to F$ is a compact linear operator, then given $\varepsilon > 0$ there is a continuous *linear* map $w \in E' \otimes F = \mathcal{P}^1_f(E;F)$ such that $||u(x) - w(x)|| < \varepsilon$ for all $x \in E$, with $||x|| \leq 1$.

What happens if neither E' nor F has the approximation property? We will prove that the above approximation is always possible if we allow the finite-rank map w to be a polynomial, i.e. an element of $\mathcal{P}_f(E;F)$. In [39] it was assumed that the space E is reflexive. We thank Prof. Charles Stegall for calling our attention to the factorization theorem of T. Figiel and W.B. Johnson that makes unnecessary the reflexivity of E.

LEMMA 4.28 (Figiel [23], Johnson [33]) *Let E and F be two real Banach spaces, and $u : E \to F$ a compact linear operator. There exists a reflexive real Banach space G and compact linear operators $v : E \to G$ and $g : G \to F$ such that $g \circ v = u$.*

THEOREM 4.29 *Let E and F be two real Banach spaces, and $u : E \to F$ a compact linear map. Then, given $\varepsilon > 0$, there is a continuous polynomial of finite type $w \in \mathcal{P}_f(E;F)$ with $w(0) = 0$ and such that*

$$||u(x) - w(x)|| < \varepsilon$$

for all $x \in E$, with $||x|| \leq 1$.

PROOF By the theorem of Figiel-Johnson there is a *reflexive* Banach space G and compact linear operators $v : E \to G$ and $g : G \to F$ such that $g \circ v = u$. Let X be a closed ball of G such that $v(x) \in X$ for all $x \in E$, $||x|| \leq 1$. Since G is reflexive, X equipped with the $\sigma(G,G')$-topology is compact. Let W be the vector subspace of $C(X;F)$ generated by $\mathcal{P}^n_f(G;F)|_X$ for all

$n \geq 1$. Then W is a polynomial algebra, separating over X, and such that, given $t \in X$, $t \neq 0$, there is $g \in W$ with $g(t) \neq 0$. Since $g : G \to F$ is a compact linear map, the restriction $g|_X$ is in $C(X;F)$. By Corollary 4.18, § 2, $g|_X$ belongs to the closure of W in $C(X;F)$. Given $\varepsilon > 0$, let $h \in W$ be such that $||g(t) - h(t)|| < \varepsilon$ for all $t \in X$. If $x \in E$, $||x|| \leq 1$, then $v(x) = t \in X$. Hence $||(g \circ v)(x) - (h \circ v)(x)|| < \varepsilon$. Let $w = h \circ v$; then $w \in \mathcal{P}_f(E;F)$ and $||u(x) - w(x)|| < \varepsilon$ for all $||x|| \leq 1$.

DEFINITION 4.30 *A mapping* $f : E \to F$ *between two Banach spaces is said to be weakly continuous if* f *is continuous from the weak topology* $\sigma(E;E')$ *in* E *to the norm topology in* F.

All $\phi \in E'$ are weakly continuous, and as a corollary all $p \in \mathcal{P}_f(E) \otimes F$ are weakly continuous too.

We shall denote by $C(E_w;F)$ the vector space of all weakly continuous maps from E into F, equipped with the topology defined by the family of seminorms

$$f \to \sup \{||f(x)||;\ x \in K\}$$

where $K \subset E$ is a weakly compact subset. If we denote by X the space $(E,\sigma(E,E'))$, then $C(E_w;F)$ with the above topology is just $C(X;F)$ with the compact-open topology.

THEOREM 4.31 *Let* E *and* F *be two real Banach spaces. Then* $\mathcal{P}_f(E) \otimes F$ *is dense in* $C(E_w;F)$.

PROOF Let $X = (E,\sigma(E,E'))$. By the remarks made after Definition 4.30, $\mathcal{P}_f(E) \otimes F$ is contained in $C(X;F)$. Since $\mathcal{P}_f(E) \otimes F$ is a polynomial algebra, which is separating and everywhere different from zero, we can apply Corollary 4.18, § 2, with $W = \mathcal{P}_f(E) \otimes F \subset C(X;F)$, to conclude that $\mathcal{P}_f(E) \otimes F$ is dense in $C(X;F) = C(E_w;F)$ in the compact-open topology.

COROLLARY 4.32 *Let* E *and* F *be two real Banach spaces and suppose that* E *is reflexive. Let* $g : E \to F$ *be a weakly continuous map and let* $r > 0$. *Given* $\varepsilon > 0$, *there is a continuous polynomial of finite type* $h \in \mathcal{P}_f(E) \otimes F$ *such that* $||g(x) - h(x)|| < \varepsilon$, *for all* $x \in E$ *with* $||x|| \leq r$.

PROOF When E is a *reflexive* Banach space, any closed ball $\{x \in E;\ ||x|| \leq r\}$ is weakly compact, and the topology of $C(E_w;F)$ can be defined by the family of seminorms

$$f \to \sup \{||f(x)||;\ ||x|| \leq r\}$$

where $r > 0$.

DEFINITION 4.33 *A mapping* $f : E \to F$ *between two Banach spaces is said to be weakly continuous on bounded sets if the restriction of* f *to any bounded subset* X *of is continuous from the relative weak topology* $\sigma(E,E')|_X$ *on* X *to the norm topology in* F.

Any weakly continuous mapping $f : E \to F$ is weakly continuous on bounded sets, but the converse is false in general, even in the case of a Hilbert space E and $F = \mathbb{R}$. (See Restrepo [51], pg. 194).

When E is a *reflexive* Banach space, we shall denote by $C(E_{wcb};F)$ the vector space of all $f : E \to F$ which are weakly continous on bounded sets, equipped with the topology defined by the seminorms

$$f \to \sup \{||f(x)||;\ x \in X\}$$

where $X \subset E$ is bounded. Since every bounded set $X \subset E$ is contained in some closed ball centered at the origin, this topology is also defined by the family of seminorms

$$f \to \sup \{||f(x)||;\ ||x|| \leq r\}$$

where $r > 0$. The following result generalizes Theorem 3 of Restrepo [53].

THEOREM 4.34 *Let* E *and* F *be two real Banach spaces and suppose that* E *is reflexive. Then* $g : E \to F$ *is weakly continuous on bounded sets, if and only if, there is a sequence* $\{p_n\}$ *of polynomials* $p_n \in \mathcal{P}_f(E) \otimes F$ *such that* $p_n \to g$ *uniformly on bounded sets.*

PROOF Let $g : E \to F$ be such that there exists a sequence $\{p_n\}$ of polynomials $p_n \in \mathcal{P}_f(E) \otimes F$ such that $p_n \to g$ uniformly on bounded sets. Let $X \subset E$ be a bounded set. Let $r > 0$ be such that $X \subset \{x \in E;\ ||x|| \leq r\} = U_r$. Let $C_b(U_r;F)$ be the Banach space of all bounded continuous mappings from U_r (equipped with the relative weak topology $\sigma(E,E')|U_r$ into the Banach space F. Since $p_n|U_r \to g|U_r$ uniformly, it follows that $g \in C_b(U_r;F)$. A fortiori, $g|X$ is continuous from the relative weak topology $\sigma(E,E')|X$ on X to the norm topology of F.

Conversely, assume that $g : E \to F$ is weakly continuous on bounded sets. Since every bounded set $X \subset E$ is contained in some closed ball $\{x \in E; ||x|| \leq n\}$, $n = 1,2,3,\ldots$, the topology of $C(E_{wcb};F)$ is metrizable and the result follows from the following.

THEOREM 4.35 *Let* E *and* F *be two real Banach spaces and suppose that* E *is reflexive. Then* $\mathcal{P}_f(E) \otimes F$ *is dense in* $C(E_{wcb};F)$.

PROOF Let $g \in C(E_{wcb};F)$ be given. For each $n = 1,2,3,\ldots$, let $U_n = \{x \in E; ||x|| \leq n\}$, equipped with the relative weak topology $\sigma(E,E')|U_n$. Then $g|U_n \in C(U_n;F)$. Let $W_n = (\mathcal{P}_f(E) \otimes F)|U_n$. Then W_n is a polynomial algebra contained in $C(U_n;F)$, which is separating and everywhere different from zero. By Corollary 4.18, § 2, W_n is dense in $C(U_n;F)$. Hence, given $\varepsilon > 0$, there is a continuous polynomial of finite type $p \in \mathcal{P}_f(E;F)$ such that

$||p(x) - g(x)|| < \varepsilon$ for all $x \in E$ with $||x|| \leq n$.

REFERENCES FOR CHAPTER 4

ARON and SCHOTTENLOHER [3]

BLATTER [4]

BIERSTEDT [5]

DE LA FUENTE [16]

FIGIEL [23]

JACOBSON [31]

JOHNSON [33]

LANG [36]

MACHADO and PROLLA [39]

NACHBIN [44]

NACHBIN, MACHADO and PROLLA [46]

PEŁCZYŃSKI [47]

PRENTER [48], [49]

PROLLA and MACHADO [52]

RESTREPO [53]

SCHAFER [58]

CHAPTER 5

WEIGHTED APPROXIMATION

§ 1 DEFINITION OF NACHBIN SPACES

Let X be a Hausdorff space. A family V of upper semicontinuous positive functions on X is said to be *directed* if given $v, w \in V$, there exists a $\lambda > 0$ and $u \in V$ such that $v(x) \leq \lambda u(x)$, $w(x) \leq \lambda u(x)$, for all $x \in X$. Any element of a directed family of upper semicontinuous positive functions on X is called a *weight on* X.

Let E be a locally convex space. A function $h: X \to E$ *vanishes at infinity* if, given $\varepsilon > 0$ and $p \in cs(E)$, the set $\{x \in X;\ p(h(x)) \geq \varepsilon\}$ is compact. Hence $p \circ h$ is upper semicontinuous, and therefore bounded on X.

DEFINITION 5.1 *Let* V *be a directed set of weights on* X. *The Nachbin space* $CV_\infty(X;E)$ *is the vector subspace of all* $f \in C(X;E)$ *such that* vf *vanishes at infinity, for each* $v \in V$, *topologyzed by the family of seminorms*

$$f \to \|f\|_{v,p} = \sup\{v(x)p(f(x));\ x \in X\}$$

where $v \in V$ *and* $p \in cs(E)$.

When $E = \mathbb{K}$, and no confusion may arise, we write simply $CV_\infty(X)$ instead of $CV_\infty(X;\mathbb{K})$.

EXAMPLE 5.2 *Let* $v : X \to \mathbb{R}$ *be defined by* $v(x) = 1$ *for all* $x \in X$, *and let* $V = \{v\}$. *Then* $CV_\infty(X;E)$ *is the vector subspace of all* $f \in C(X;E)$ *that vanish at infinity. This space is usually denoted by* $C_0(X;E)$. *Its topology* σ *is the topology of uniform convergence on* X.

The vector subspace of all $f \in C(X;E)$ such that the support of f is compact will be denoted by $K(X;E)$. Obviously, $K(X;E) \subset C_0(X;E)$. If X is compact, $K(X;E) = C_0(X;E) = C(X;E)$.

If $p \in cs(E)$ and $K \subset X$ is a compact subset, then

$$\sup\{p(f(x));\ x \in K\} \leq \sup\{p(f(x));\ x \in X\}$$

for all $f \in C_0(X;E)$. This shows that the topology of uniform convergence on X is stronger than the compact-open topology κ induced by $C(X;E)$ on $C_0(X;E)$.

EXAMPLE 5.3 *Let X be a locally compact Hausdorff space. Consider the directed family* $V = \{\phi \in C_0(X;\mathbb{R});\ \phi \geq 0\}$. *Then* $CV_\infty(X;E) = C_b(X;E)$ *as vector spaces and the topology defined by the family of seminorms*

$$f \to \sup\{\phi(x)p(f(x));\ x \in X\} = ||f||_{\phi,p}$$

on $C_b(X;E)$ *is called the strict topology and it is denoted by* β. *(see Buck* [11]*).*

The strict topology β is stronger than the compact-open topology induced on $C_b(X;E)$ by $C(X;E)$; on the other hand, β is weaker than the topology σ of uniform convergence on X.

EXAMPLE 5.4 *Let V be the set of all characteristic functions of compact subsets* $K \subset X$. *Then the Nachbin space* $CV_\infty(X;E)$ *is just* $C(X;E)$ *endowed with the compact-open topology.*

§ 2 THE BERNSTEIN-NACHBIN APPROXIMATION PROBLEM

Let $W \subset CV_\infty(X;E)$ be a vector subspace which is an A-module, where $A \subset C(X;\mathbb{K})$ is a subalgebra.

The *Bernstein-Nachbin approximation problem* consists in asking for a description of the closure of W in $CV_\infty(X;E)$.

Let P be a closed, pairwise disjoint covering of X. We say that W is *P-localizable* in $CV_\infty(X;E)$ if the closure of W in $CV_\infty(X;E)$ consists of those $f \in CV_\infty(X;E)$ such that, given any $S \in P$, any $v \in V$, any $p \in cs(E)$, and any $\varepsilon > 0$, there is some $g \in W$ such that

$$v(x)\ p(f(x) - g(x)) < \varepsilon$$

for all $x \in S$.

The *strict Bernstein-Nachbin approximation problem* consists in asking for necessary and sufficient conditions for an A-module W to be P-localizable, when P is the set P_A of all equivalence classes $Y \subset X$ modulo $X|A$.

In [46], the sufficient conditions for localizability established by Nachbin (see e.g. Nachbin [43]) were extended to the context of vector-fibrations, and a fortiori to vector-valued functions, in the case of modules over *real* or *self-adjoint complex* algebras. In [40], the results of [46] were extended to the *general complex* case in the same way that Bishop's Theorem generalizes the Stone-Weierstrass Theorem.

Before stating Definition 5.5, we recall that $\mathcal{P}(\mathbb{R}^n)$ denotes the algebra of all $\mathbb{R}$-valued polynomials on $\mathbb{R}^n$.

DEFINITION 5.5 *Let* ω *be a weight on* $\mathbb{R}^n$. *The weight* ω *is said to be rapidly decreasing at infinity when* $\mathcal{P}(\mathbb{R}^n) \subset C\omega_\infty(\mathbb{R}^n)$.

If ω is a rapidly decreasing weight on $\mathbb{R}^n$, then ω is called a *fundamental weight* in the sense of Serge Bernstein, if $\mathcal{P}(\mathbb{R}^n)$ is dense in $C\omega_\infty(\mathbb{R}^n)$. We shall denote by Ω_n the set of all fundamental weights on $\mathbb{R}^n$.

We denote by Ω_n^s the subset of Ω_n consisting of those $\omega \in \Omega_n$ which are *symmetric* in the sense $\omega(t) = \omega(|t|)$, for all $t \in \mathbb{R}^n$, where $|t| = (|t_1|,\dots,|t_n|)$ if $t = (t_1,\dots,t_n)$.

We denote by Γ_1 the subset of Ω_1 consisting of those $\gamma \in \Omega_1$ such that $\gamma^k \in \Omega_1$ for any real number $k > 0$. Let then $\Gamma_1^s = \Gamma_1 \cap \Omega_1^s$.

We notice the inclusion $\Omega_n^d \subset \Omega_n^s$ and similarly $\Gamma_1^d \subset \Gamma_1^s$. Here Ω_n^d denotes the subset of all $\omega \in \Omega_n$ such that $|u| \leq |t|$ implies $\omega(u) \geq \omega(t)$ for all $u,t \in \mathbb{R}^n$ and then $\Gamma_1^d = \Gamma_1 \cap \Omega_1^d$.

DEFINITION 5.6 *Let* P *be a closed, pairwise disjoint covering of* X. *We say that* W *is sharply* P*-localizable in* $CV_\infty(X;E)$ *if, given* $f \in CV_\infty(X;E)$, $v \in V$ *and* $p \in cs(E)$, *there is some* $S \in P$ *such that*

$$\inf\{||f-g||_{v,p};\ g \in W\} = \inf\{||f|S - g|S||_{v,p};\ g \in W\}.$$

DEFINITION 5.7 *For each* $v \in V$, $p \in cs(E)$, *and* $\delta > 0$, *we denote by* $L(W;v,p,\delta)$ *the set of all* $f \in CV_\infty(X;E)$ *such that, for each equivalence class* $Y \subset X$ *(mod. A) there is* $g \in W$ *such that* $||f|Y - g|Y||_{v,p} < \delta$.

In our next definition, $\mathfrak{S}$ is the class of all ordinal numbers whose cardinal numbers are less or equal than $2^{|X|}$, where $|X|$ is the cardinal number of X. For each $\sigma \in \mathfrak{S}$, P_σ is the closed, pairwise disjoint covering of X defined in §8, Chapter 1.

DEFINITION 5.8 *We say that the A-module W is sharply localizable under A in* $CV_\infty(X;E)$ *if, given* $f \in CV_\infty(X;E)$, $v \in V$, *and* $p \in cs(E)$, *for each* $\sigma \in \mathfrak{S}$ *there exists an element* $S_\sigma \in P_\sigma$ *such that:*

(a) $S_\sigma \subset S_\tau$ *for all* $\tau < \sigma$,

(b) $\inf \{||f-g||_{v,p}; g \in W\} = \inf\{||f|S_\sigma - g|S_\sigma||_{v,p};$

$g \in W\}$.

DEFINITION 5.9 *We say that a subset* $G(A) \subset A$ *is a set of generators for* A, *if the subalgebra over* $\mathbb{K}$ *generated by* G(A) *is dense in* A *for the compact-open topology of* $C(X;\mathbb{K})$; *and we say that a set of generators* $G(A) \subset A$ *is a strong set of generators if, for any* $\sigma \in \mathfrak{S}$ *and any* $S \in P_\sigma$, *the set* $A_S \cap G(A)$ *is a set of generators for the algebra* A_S *(Recall that* $A_S = \{a \in A;\ a|S$ *is real-valued*}*). For example, the whole algebra* A *is a strong set of generators for* A. *Also, if the algebra* A *has a set of generators* G(A) *consisting only of real-valued functions, then* G(A) *is a strong set of generators for the algebra* A.

Similarly, a subset $G(W) \subset W$ is a *set of generators for* W if the A-submodule of W generated by G(W) is dense in for the topology of $CV_\infty(X;E)$. Let us call $G(W)^*$ the *real* linear span of G(W).

LEMMA 5.10 *Let* $A \subset C_b(X;\mathbb{R})$ *be a subalgebra containing the constants. For each equivalence class* $Y \subset X$ *modulo* $X|A$, *let there be given a compact set* $K_Y \subset X$, *disjoint from* Y. *Then, there exist equivalence classes* $Y_1,\ldots,Y_n \subset X$ *modulo* $X|A$ *such that to each* $\delta > 0$, *there correspond functions* $a_1,\ldots,a_n$ *in* A *satisfying the following properties:*

(a) $0 \leq a_i \leq 1$, $i = 1,\ldots,n$;

(b) $0 \leq a_i(t) < \delta$, *for* $t \in K_{Y_i}$, $i = 1,\ldots,n$;

(c) $a_1 +\ldots+ a_n = 1$ *on* X.

PROOF Let P_A be the set of all equivalence classes $Y \subset X$ modulo $X|A$. Select one element Y_1 in P_A, and let P be the collection of all elements $Y \in P_A$ such that the intersection $Y \cap K_{Y_1}$ is non-empty. Choose a real number $0 < \varepsilon < 1 - \varepsilon$. For each $Y \in P_A$, there is $b_Y \in A$ given by Lemma 1.3, § 3, Chapter 1, with $U = X\setminus K_Y$. Let $B_Y = \{x \in X;\ b_Y(x) > 1 - \varepsilon\}$. Clearly, $Y \subset B_Y$, so that the collection $\{B_Y;\ Y \in P\}$ is an open covering of the compact subset $K_{Y_1} \subset X$. By compactness, there are equivalence classes $Y_2,\ldots,Y_n$ in P such that $K_{Y_1} \subset B_2 \cup\ldots\cup B_n$, where we have written $B_i = B_Y$ for $Y = Y_i$, $i = 2,\ldots,n$. For each index $i = 2,\ldots,n$, there is a polynomial $p_i : \mathbb{R} \to \mathbb{R}$ such that

(1) $p_i(1) = 1$;

(2) $0 \leq p_i(t) \leq 1$, $t \in [0,1]$;

(3) $0 \leq p_i(t) < \delta$, $t \in [0,\varepsilon]$;

(4) $1 - \delta < p_i(t) \leq 1$, $t \in [1 - \varepsilon, 1]$.

Indeed, apply Lemma 1.4, § 3, Chapter 1, to get such polynomials. Consider $g_i = p_i(b_i)$, where $b_i = b_Y$, for $Y = Y_i$, $i = 2,\ldots,n$. Then $g_i \in A$, $i = 2,\ldots,n$. Define

$$a_2 = g_2$$

$$a_3 = (1 - g_2)g_3$$

$$\cdots\cdots\cdots\cdots\cdots$$

$$a_n = (1-g_2)(1-g_3)\ldots(1-g_{n-1})g_n.$$

For $i = 2,\dots,n$, it is easily seen that $a_i \in A$; $0 \leq a_i \leq 1$; and $a_i(x) \leq g_i(x) < \delta$ for all $x \in K_i$, where $K_i = K_Y$ with $Y = Y_i$. Moreover, by induction, we see that

$$a_2 + \dots + a_n = 1 - (1 - g_2)(1 - g_3)\dots(1 - g_n).$$

Let $a_1 = 1 - (a_2 + \dots + a_n)$. Then $a_1 \in A$; $0 \leq a_1 \leq 1$, and $a_1 + \dots + a_n = 1$ on X, which proves (a) and (c) of the statement. To prove (b), it only remains to prove that $a_1(x) < \delta$ for all $x \in K_{Y_1}$. Now $K_{Y_1} \subset B_2 \cup \dots \cup B_n$, so that $x \in B_j$, for some $j = 2,\dots,n$, and therefore $1 - g_j(x) < \delta$, and so

$$a_1(x) = (1 - g_j(x)) \prod_{\substack{i=2 \\ i \neq j}}^{n} (1 - g_i(x)) < \delta .$$

THEOREM 5.11 *Suppose that there exist sets of generators* G(A) *and* G(W), *for* A *and* W *respectively, such that:*

(1) G(A) *consists only of real valued functions;*

(2) *given any* $v \in V$, $a_1,\dots,a_n \in G(A)$, $g \in G(W)$, *and* $p \in cs(E)$, *there are* $a_{n+1},\dots,a_N \in G(A)$, *with* $N \geq n$, *and* $\omega \in \Omega_N$ *such that* $v(x)p(g(x)) \leq \omega(a_1(x),\dots,a_n(x),\dots,a_N(x))$ *for all* $x \in X$.

Then W *is sharply localizable under* A *in* $CV_\infty(X;E)$.

We first remark that, since G(A) consists only of real valued functions, $\rho = 2$ and $P_2 = P_A$, where P_A is the closed, pairwise disjoint partition of X into equivalence classes modulo $X|A$. Hence, all that we have to prove is that W is sharply P_A-localizable in $CV_\infty(X;E)$. The proof will be partitioned into several lemmas, and to state them we need a preliminary definition.

DEFINITION 5.12 *Let us call* B *the subalgebra of* $C_b(X;R)$ *of all functions of the form* $q(a_1,\dots,a_n)$, *where* $n \geq 1$, $a_1,\dots,a_n \in G(A)$, *and* $q \in C_b(\mathbb{R}^n;\mathbb{R})$ *are arbitrary.*

LEMMA 5.13 *Assume that* $G(A)$ *consists only of real valued functions. Let* $f \in L(W;v,p,\lambda)$. *Then, for each* $\varepsilon > 0$, *there exist* $b_1,\ldots,b_m \in B$, *and* $g_1,\ldots,g_m \in G(W)$ *such that*

$$\left\|f - \sum_{i=1}^{m} b_i g_i\right\|_{v,p} < \lambda + \varepsilon.$$

PROOF For each $Y \in P_A$, there exists $w_Y \in G(W)^*$ such that $v(x)\,p(f(x) - w_Y(x)) < \lambda + \varepsilon/2$, for all $x \in Y$. Let us define $K_Y = \{t \in X;\ v(t)p(f(t) - w_Y(t)) \geq \lambda + \varepsilon/2\}$. Then K_Y is compact and disjoint from Y. Since the equivalence relations $X|A$ and $X|B$ are the same, we may apply Lemma 5.10 for the algebra B. Hence, there exist equivalence classes $Y_1,\ldots,Y_n \in P_A$ such that to each $\delta > 0$, there correspond $h_1,\ldots,h_n \in B$ with $0 \leq h_i \leq 1; 0 \leq h_i(x) < \delta$ for $x \in K_i$, where $K_i = K_{Y_i}$ for $i = 1,\ldots,n$. Moreover, $h_1 + \ldots + h_n = 1$ on X. Let us choose $\delta > 0$ such that $nM\delta < \varepsilon/2$, where $M = \max\{\|f - w_i\|_{v,p};\ i = 1,\ldots,n\}$, and $w_i = w_Y$ with $Y = Y_i$ for $i = 1,\ldots,n$. Let $w = h_1w_1 + \ldots + h_nw_n$. We claim that $v(x)p(f(x) - w(x)) < \lambda + \varepsilon$, for all $x \in X$. Indeed,

$$v(x)p(f(x) - w(x)) \leq \sum_{i=1}^{n} h_i(x)v(x)p(f(x) - w_i(x)),$$

for all $x \in X$. Now, if $x \in K_i$ then $h_i(x) < \delta$, and therefore

$$h_i(x)v(x)p(f(x) - w_i(x)) < \delta \cdot \|f - w_i\|_{v,p} \leq \delta M;$$

on the other hand, if $x \notin K_i$, then the following estimate is true:

$$h_i(x)v(x)p(f(x) - w_i(x)) \leq h_i(x)(\lambda + \varepsilon/2).$$

Combining both estimates, we get

$$v(x)p(f(x) - w(x)) < nM\delta + (\lambda + \varepsilon/2)(h_1(x) + \ldots + h_n(x)) < \lambda + \varepsilon.$$

Since each $w_i \in G(W)^*$, there exist $b_1,\ldots,b_m \in B$ and $g_1,\ldots,g_m \in G(W)$ such that $w = b_1g_1 + \ldots + b_mg_m$.

LEMMA 5.14 *Suppose that the hypothesis of Theorem* 5.11 *are satisfied. Given* $v \in V$, $p \in cs(E)$, $b \in B$, $g \in G(W)$ *and* $\delta > 0$, *there is* $w \in W$ *such that* $\|w - bg\|_{v,p} < \delta$.

PROOF Suppose that $b = q(a_1,...,a_n)$. Given $v \in V$, $p \in cs(E)$, and $g \in G(W)$ there are $a_{n+1},...,a_N \in G(A)$, where $N \geq n$, and $\omega \in \Omega_N$ such that $v(x)p(g(x)) \leq \omega(a_1(x),...,a_n(x),...,a_N(x))$ for all $x \in X$. Define $r \in C_b(\mathbb{R}^n;\mathbb{R})$ by setting $r(t) = a(t_1,...,t_n)$ for all $t = (t_1,...,t_n,...,t_N) \in \mathbb{R}^N$. By hypothesis $\omega \in \Omega_N$; hence $C_b(\mathbb{R}^N;\mathbb{R})$ is contained in $C\omega_\infty(\mathbb{R}^N;\mathbb{R})$ and $\mathcal{P}(\mathbb{R}^N)$ is dense in $C\omega_\infty(\mathbb{R}^N;\mathbb{R})$. Given $\delta > 0$, we can find a real polynomial $q \in \mathcal{P}(\mathbb{R}^N)$ such that $||q - r||_\omega < \delta$. From this it follows that $||w - bg||_{v,p} < \delta$, where $w = q(a_1,...,a_n,...,a_N)g \in AW \subset W$.

LEMMA 5.15 *Suppose that the hypothesis of Theorem* 5.11 *are satisfied. Then, for each* $f \in CV_\infty(X;E)$, $v \in V$, *and* $p \in cs(E)$, *we have* $d = \inf\{||f-g||_{v,p};\ g \in W\} = \sup\{\inf\{||f|Y-g|Y||_{v,p}; g \in W\};\ Y \in P_A\}$.

PROOF Clearly, $c \leq d$, where we have defined
$c = \sup\{\inf\{||f|Y - g|Y||_{v,p};\ g \in W\};\ Y \in P_A\}$. To prove the reverse inequality, let $\varepsilon > 0$. For each $Y \in P_A$, there exists $g_Y \in W$ such that $v(x)p(f(x) - g_Y(x)) < c + \varepsilon/3$ for all $x \in Y$. Therefore, $f \in L(W;v,p,c + \varepsilon/3)$. By Lemma 5.13, applied with $\lambda = c + \varepsilon/3$ and $\varepsilon/3$, there exist $b_1,...,b_m \in B$ and $g_1,...,g_m \in G(W)$ such that

$$||f - \sum_{i=1}^{m} b_i g_i||_{v,p} < (c + \varepsilon/3) + \varepsilon/3.$$

By Lemma 5.14, applied with $\delta = \varepsilon/3m$, there are $w_1,...,w_m \in W$ such that $||w_i - b_i g_i||_{v,p} < \varepsilon/3m$. From this it follows that $||f - g||_{v,p} < c + \varepsilon$, where $g = w_1 + ... + w_m$. Since $g \in W$, $d < c + \varepsilon$. Since $\varepsilon > 0$ was arbitrary, $d \leq c$, as desired.

PROOF OF THEOREM 5.11 Let $f \in CV_\infty(X;E)$, $v \in V$, and $p \in cs(E)$ be given. Let Z be the quotient space of X by the equivalence relation $X|A$, and let $\pi : X \to Z$ be the quotient map. By Lemma 1 of [46] the map

$$z \in Z \to ||f|\pi^{-1}(z) - g|\pi^{-1}(z)||_{v,p}$$

is upper semicontinuous and null at infinity on Z, for each $g \in W$. Hence the map defined by

$$h(z) = \inf \{ \| f|\pi^{-1}(z) - g|\pi^{-1}(z) \|_{v,p};\ g \in W\}$$

for all $z \in Z$, is upper semicontinuous and null at infinity on Z too. Therefore h attains its supremum on Z at some point z. Consider the equivalence class $Y = \pi^{-1}(z)$ modulo $X|A$. On the other hand, the supremum of the map h is by Lemma 5.15 equal to d. Thus, we have found an equivalence class $Y \subset X$ modulo $X|A$ such that

$$\inf \{ \|f-g\|_{v,p};\ g \in W\} = \inf \{ \|f|Y - g|Y\|_{v,p};\ g \in W\}.$$

By the remark made before Definition 5.12 the module W is sharply localizable under A in $CV_\infty(X;E)$.

THEOREM 5.16 *Suppose that there exist sets of generators* G(A) *and* G(W), *for* A *and* W *respectively, such that:*

(1) G(A) *is a strong set of generators for* A;

(2) *given any* $v \in V$, $p \in cs(E)$, $a_1,\dots,a_n \in G(A)$ *and* $g \in G(W)$, *there exists* $\omega \in \Omega_n^s$ *such that* $v(x)p(g(x)) \leq \omega(|a_1(x)|,\dots,|a_n(x)|)$ *for all* $x \in X$.

Then W *is sharply localizable under* A *in* $CV_\infty(X;E)$.

PROOF Let $f \in CV_\infty(X;E)$, $v \in V$, and $p \in cs(E)$ be given. Let $\sigma \in \mathfrak{S}$. Assume that for each $\tau > \sigma$ we have found an element $S_\tau \in P_\tau$ such that

(a) $S_\tau \subset S_\mu$ for all $\mu < \tau$;

(b) $\inf\{\|f-g\|_{v,p};\ g \in W\} = \inf\{\|f|S_\tau - g|S_\tau\|_{v,p};\ g \in W\}$.

FIRST CASE. $\sigma = \tau + 1$ for some $\tau \in \mathfrak{S}$. By the induction hypothesis there is $S_\tau \in P_\tau$ such that (a) and (b) are true. Let A_τ be the subalgebra of all $a \in A$ such that $a|S_\tau$ is real-valued. By Theorem 5.11 applied to the algebra $A_\tau|S_\tau$ and the module $W|S_\tau$ there is a set $S_\sigma \in P_\sigma = P_{\tau+1}$ such that

$$\inf_{g \in W} ||f|S - g|S||_{v,p} = \inf_{g \in W} ||f|S - g|S||_{v,p}$$

On the other hand, $S_\sigma \subset S_\tau$, by construction.

SECOND CASE. The ordinal $\sigma \in \mathfrak{S}$ has no predecessor. Define $S_\sigma = \cap\{S_\tau;\ \tau < \sigma\}$. Then $S_\sigma \in P_\sigma$ and $S_\sigma \subset S_\tau$ for all $\tau < \sigma$. Assume that $\inf\{||f|S_\sigma - g|S_\sigma||_{v,p};\ g \in W\} < d$, where we have defined $d = \inf\{||f - g||_{v,p};\ g \in W\}$. (The case $d = 0$ is trivial). There exists $g \in W$ such that $||f|S_\sigma - g|S_\sigma||_{v,p} < d$. Let $U \subset X$ be the open set $\{t \in X;\ v(t)p(f(t) - g(t)) < d\}$. Then the complement of U in X is compact, and $S_\sigma \subset U$. By compactness, there exist $\tau_1 < \ldots < \tau_n < \sigma$ such that $X\backslash U \subset (X\backslash S_1) \cup \ldots \cup (X\backslash S_n)$, where $S_i = S_\tau$ with $\tau_i = \tau$. However, since $S_n \subset \ldots \subset S_1$, it follows that $S_n \subset U$, a contradiction to (b), because $\tau_n < \sigma$.

§ 3 SUFFICIENT CONDITIONS FOR SHARP LOCALIZABILITY

THEOREM 5.17 *Suppose that there exist sets of generators* G(A) *and* G(W), *for* A *and* W *respectively, such that:*

(1) G(A) *is a strong set of generators for* A;

(2) *given any* $v \in V$, $a \in G(A)$, $g \in G(W)$, *and* $p \in cs(E)$, *there exists* $\gamma \in \Gamma_1^s$ *such that* $v(x)p(g(x)) \leq \gamma(|a(x)|)$ *for all* $x \in X$.

Then W *is sharply localizable under* A *in* $CV_\infty(X;E)$.

PROOF Given any $v \in V$, $a_1,\ldots,a_n \in G(A)$, $g \in G(W)$, and $p \in cs(E)$, there are $\gamma_i \in \Gamma_1^s$ such that $v(x)p(g(x)) \leq \gamma_i(|a_i(x)|)$ for all $x \in X$, $i = 1,\ldots,n$. Define ω on R^n by $\omega(t) = [\gamma_1(t_1)\ldots\gamma_n(t_n)]^{1/n}$ for all $t = (t_1,\ldots,t_n)$. Then $\omega \in \Omega_n$ by Lemma 1, § 27, [43]. Obviously, $\omega(t) = \omega(|t|)$ for all $t \in R^n$. Hence, $\omega \in \Omega_n^s$. By Theorem 5.16, W is sharply localizable under A in $CV_\infty(X;E)$.

THEOREM 5.18 *(Analytic criterion) Suppose that there exist sets of generators* $G(A)$ *and* $G(W)$ *such that:*

(1) $G(A)$ *is a strong set of generators for* A;

(2) *given any* $v \in V$, $a \in G(A)$, $g \in G(W)$, *and* $p \in cs(E)$, *there are constants* $M > 0$ *and* $m > 0$ *such that* $v(x)p(g(x)) \leq M\, e^{-m|a(x)|}$ *for all* $x \in X$.

Then W *is sharply localizable under* A *in* $CV_\infty(X;E)$.

PROOF The function $\gamma(t) = M\, e^{-m|t|}$ defined for all $t \in \mathbb{R}$, belongs to Γ_1^s by Lemma 2, § 28 of [43]. It remains to apply Theorem 5.17 above.

THEOREM 5.19 *(Quasi-analytic criterion) Suppose that there exist sets of generators* $G(A)$ *and* $G(W)$ *such that:*

(1) $G(A)$ *is a strong set of generators for* A;

(2) *given any* $v \in V$, $a \in G(A)$, $g \in G(W)$, *and* $p \in cs(E)$, *we have* $\sum_{m=1}^{\infty} (M_m)^{-1/m} = +\infty$, *where*

$$M_m = \|a^m g\|_{v,p} \quad \textit{for} \quad m = 0,1,2,\ldots .$$

Then W *is sharply localizable under* A *in* $CV_\infty(X;E)$.

PROOF Define γ on $\mathbb{R}$ as in the proof of Theorem 9, [46], and then apply Theorem 5.17, above.

THEOREM 5.20 *(Bounded case) Suppose that there exist sets of generators* $G(A)$ *and* $G(W)$, *for* A *and* W *respectively, such that*

(1) $G(A)$ *is a strong set of generators for* A;

(2) *given any* $v \in V$, $a \in G(A)$, $g \in G(W)$, *and* $p \in cs(E)$, *the function* a *is bounded on the support of* $v\, p(g)$.

Then W *is sharply localizable under* A *in* $CV_\infty(X;E)$.

PROOF Let $v \in V$, $a \in G(A)$, $p \in cs(E)$ and $g \in G(W)$ be given. Let $m > \sup\{|a(x)|;\ x \in S\}$, where S is the support of the function $vp(g)$; and let $M > \|g\|_{v,p}$. If γ is the characteristic function of the interval $[-m,m] \subset \mathbb{R}$ times the constant M, then $\gamma \in \Gamma_1^s$ and $v(x)p(g(x)) \leq \gamma(|a(x)|)$ for all $x \in X$. It remains to apply Theorem 5.17.

REMARK The above Theorem 5.20 generalizes Theorem 4, § 2 of Kleinstück [35], which in turn was a generalization of Theorem 4.5 of Prolla [51] and of the result of Summers [64].

COROLLARY 5.21 *Let* $W \subset CV_\infty(X;E)$ *be an A-module. Suppose that every* $a \in A$ *is bounded on the support of every* $v \in V$. *Then* W *is sharply localizable under* A *in* $CV_\infty(X;E)$.

PROOF The set A is a strong set of generators for A. Since the support of v contains the support of $x \to v(x)p(g(x))$ for any continuous seminorm $p \in cs(E)$ and any $g \in W$, we may apply Theorem 5.20 with $G(A) = A$ and $G(W) = W$.

COROLLARY 5.22 *(Kleinstuck [35]). Assume the hypothesis of Corollary* 5.21. *Then for every* $f \in CV_\infty(X;E)$, f *belongs to the closure of* W *in* $CV_\infty(X;E)$ *if, and only if, given any* $v \in V$, $p \in cs(E)$, $\varepsilon > 0$, *and* $K \in \mathcal{K}_A$ *there exists* $g \in W$ *such that* $v(x)p(f(x) - g(x)) < \varepsilon$ *for all* $x \in K$.

PROOF By Corollary 5.21, W is sharply localizable under A in $CV_\infty(X;E)$. Since $P_\rho = \mathcal{K}_A$, W is sharply $\mathcal{K}_A$-localizable, i.e., given $f \in CV_\infty(X;E)$, $v \in V$, $p \in cs(E)$, there is some maximal antisymmetric set $K \in \mathcal{K}_A$ such that

$$\inf\{||f-g||_{v,p};\ g \in W\} = \inf\{||f|K - g|K||_{v,p};\ g \in W\}.$$

This formula generalizes that obtained by Glicksberg in the case of Bishop's Theorem (see [26]), and from it there follows the desired conclusion.

§ 4 COMPLETENESS OF NACHBIN SPACES

$F(X,E)$ is the vector space of all mappings from X into E, and $B(X,E)$ is the vector subspace of all mappings f from X into E such that $f(X)$ is a bounded subset of E, and $B_o(X,E)$ is the vector subspace of $B(X,E)$ consisting of those bounded mappings f from X into E that vanish at infinity, i.e., those $f \in B(X,E)$ such that, given any continuous seminorm p on E and any $\varepsilon > 0$, there is a compact subset $K \subset X$ such that

$p(f(x)) < \varepsilon$ for every $x \in X$ outside of K. The vector subspace $C(X,E) \cap B(X,E)$ is $C_b(X,E)$, and $C(X,E) \cap B_o(X,E)$ is $C_o(X,E)$. Finally $K(X,E)$ will denote the subspace of $C(X,E)$ consisting of those functions that are identically equal to 0 in E outside of some compact subset of X. The corresponding spaces for $E = \mathbb{R}$ or $\mathbb{C}$ are written omitting E.

If V is a directed set of weights on X, the vector spaces of all $f \in F(X,E)$ such that $vf \in B(X,E)$, for any $v \in V$, is denoted by $FV_b(X,E)$. On $FV_b(X,E)$ we shall consider the locally convex topology determined by the family of seminorms $||f||_{v,p} = \sup \{v(x)p(f(x));\ x \in X\}$ when v ranges over V and p ranges over the set of all continuous seminorms on E. This topology will be denoted by ω_V, and the space $FV_b(X,E)$ endowed with ω_V is called a weighted space of vector-valued functions. It has a basis of closed absolutely convex neighborhoods of the origin of the form $N_{v,p} = \{f \in FV_b(X,E);\ ||f||_{v,p} \leq 1\}$. The vector subspace of $FV_b(X,E)$ consisting of those $f \in F(X,E)$ such that $vf \in B_o(X,E)$, for any $v \in V$, is denoted by $FV_\infty(X,E)$ and it is endowed with the topology induced by $FV_b(X,E)$. $FV_\infty(X,E)$ is a closed subspace of $FV_b(X,E)$. We shall always assume that $FV_b(X,E)$ is a Hausdorff space. This is clearly the case if V is everywhere different from zero on X, i.e., if for every $x \in X$ there is a $v \in V$ such that $v(x) > 0$ (one says then that $V > 0$ on X).

EXAMPLE 5.23 For any subset $S \subset X$, the characteristic function of S will be denoted by χ_S. Let $\chi_f(X) = \{\lambda\chi_F;\ \lambda \geq 0$ and $F \subset X, F$ finite$\}$. Then $V = \chi_f(X)$ is directed set of weights on X and $FV_b(X,E) = FV_\infty(X,E) = F(X,E)$. The topology ω_V in this case is the topology ω of pointwise convergence.

EXAMPLE 5.24 Let $\chi_c(X) = \{\lambda\chi_K;\ \lambda \geq 0$ and $K \subset X$, K compact$\}$. Then $V = \chi_c(X)$ is a directed set of weights on X and $FV_b(X,E) = FV_\infty(X,E)$. The topology ω_V in this case is the topology κ of compact convergence, determined by the family of seminorms $||f||_{K,p} = \sup \{p(f(x));\ x \in K\}$, when K ranges over all compact subsets of X, and p ranges over all continuous seminorms on E.

EXAMPLE 5.25 Let $K^+(X)$ be the set of all constant ≥ 0 functions on X. Then $V = K^+(X)$ is a directed set of weights on X and $FV_b(X,E) = B(X,E)$, $FV_\infty(X,E) = B_0(X,E)$. The topology ω_V in this case is the topology σ of uniform convergence, determined by the family of seminorms $||f||_p = \sup\{p(f(x));\ x \in X\}$, when p ranges over all continuous seminorms on E.

If V is a directed set of weights on X, clearly $CV_b(X,E)$ and $CV_\infty(X,E)$ are the intersections $FV_b(X,E) \cap C(X,E)$ and $FV_\infty(X,E) \cap C(X,E)$ respectively. Those spaces are equipped with the topology induced by ω_V. When $V = \chi_f(X)$, $CV_b(X,E) = CV_\infty(X,E) = C(X,E)$ equipped with the topology ω of pointwise convergence. When $V = \chi_c(X)$, $CV_b(X,E) = CV_\infty(X,E) = C(X,E)$ equipped with the topology κ of compact convergence. When $V = K^+(X)$, $CV_b(X,E) = C_b(X,E)$ and $CV_\infty(X,E) = C_0(X,E)$ both equipped with the topology σ of uniform convergence.

DEFINITION 5.26 (Summers) *If* U *an* V *are directed sets of weights on* X *and for every* $u \in U$ *there is a* $v \in V$ *such that* $u \leq v$, *then we write* $U \leq V$. *In case* $U \leq V$ *and* $V \leq U$, *we write* $U \approx V$.

Let U be a directed set of weights on X and $\phi: X \to X$ a mapping. Then, for every V on X such that $U \leq V \circ \phi$ the mapping $f \to f \circ \phi$ is a continuous linear mapping from $FV_b(X,E)$ into $FU_b(X,E)$. Indeed, given $f \in FV_b(X,E)$ and $u \in U$, choose $v \in V$ such that $u \leq v \circ \phi$. Then, for any continuous seminorm p on E, we have

$$||f \circ \phi||_{u,p} \leq \sup\{v(\phi(x))p(f(\phi(x)));\ x \in X\} \leq ||f||_{v,p}.$$

Hence to get a continuous linear mapping from the space $FV_\infty(X,E)$ into $FU_\infty(X,E)$ it is sufficient to assume that for every compact subset $K \subset X$ the intersection of $\phi^{-1}(K)$ with the support of any $u \in U$ is compact. Indeed, if $f \in FV_\infty(X,E)$ and $u \in U$, choose $v \in V$ such that $u \leq v \circ \phi$. We know that $f \in FU_b(X,E)$. Now, given any continuous seminorm p on E and $\varepsilon > 0$, there exists a compact subset $K \subset X$ such that $v(x)p(f(x)) < \varepsilon$ for all $x \notin K$. Let K' be the intersection of $\phi^{-1}(K)$ with the support of u. Let $x \notin K'$. Then either $x \notin \phi^{-1}(K)$ or x is not in the support of u. Suppose first that $x \notin \phi^{-1}(K)$. Then $\phi(x) \notin K$, hence $v(\phi(x))p(f(\phi(x))) < \varepsilon$. This implies that

$u(x)p(f(\phi(x))) < \varepsilon$. If x is not in the support of u then $u(x)=0$ and again $u(x)p(f(\phi(x))) < \varepsilon$. Hence $u(f \circ \phi) \in B_o(X,E)$. Since $u \in U$ was arbitrary, $f \circ \phi \in FV_\infty(X,E)$. We have thus proved the following.

PROPOSITION 5.27 *Let* U *be a directed of weights on* X *and* $\phi : X \to X$ *be a mapping such that for every compact subset* $K \subset X$, *the intersection of* $\phi^{-1}(K)$ *with the support of any* $u \in U$ *is compact. Then, for every directed set of weights* V *on* X *such that* $U \leq V \circ \phi$, *the mapping* $f \to f \circ \phi$ *is a continuous linear mapping from* $FV_\infty(X,E)$ *into* $FU_\infty(X,E)$.

PROPOSITION 5.28 *If* U *and* V *are directed sets of weights on* X *with* $U \leq V$, *then*

(1) $FV_b(X,E) \subset FU_b(X,E)$;

(2) $FV_\infty(X,E) \subset FU_\infty(X,E)$;

(3) *the topology induced on* $FV_b(X,E)$ *by* ω_U *is weaker than* ω_V.

PROOF In Proposition 5.27 take ϕ equal to the identity map on X.

COROLLARY 5.29 *If* U *and* V *are two directed sets of weights on* X *with* $U \approx V$, *then* $FU_b(X,E) = FV_b(X,E)$ *and* $FU_\infty(X,E) = FV_\infty(X,E)$ *as topological vector spaces.*

Let V be a directed set of weights on X such that the weights $v \in V$ vanish at infinity and $\chi_c(X) \leq V$. Then $B(X,E) \subset FV_\infty(X,E)$ and the topology induced on $B(X,E)$ by ω_V enjoys some of the properties of the strict topology β. Namely, we have the following result.

THEOREM 5.30 *Let* V *be a directed set of weights on* X *such that* $\chi_c(X) \leq V$ *and all weights* $v \in V$ *vanish at infinity. Then on* $B(X,E)$:

(1) $\kappa \leq \omega_V \leq \sigma$;

(2) *every* σ*-bounded subset is* ω_V*-bounded and every* ω_V*-bounded subset is* κ*-bounded;*

(3) *ω_V and κ agree on σ-bounded subsets.*

PROOF Since $V \leq K^+(X)$, (1) follows from Proposition 5.28, while (2) is an immediate consequence of (1). To prove (3) let S be a σ-bounded subset of $B(X,E)$ and let $f \in B(X,E)$ be in the κ-closure of S. Let p be a continuous seminorm on E, $v \in V$ and $\varepsilon > 0$. Let $M > 0$ be such that $||g||_p \leq M$ for all $g \in S$, and $K \subset X$ be a compact subset of X such that $v(x) < \varepsilon/2(M + ||f||_p)$ for all $x \notin K$. Let $g \in S$ be such that $||f-g||_{K,p} < \varepsilon/2||v||$, where $||v|| = \sup\{v(x);\ x \in X\}$. Then

$$
\begin{aligned}
||f-g||_{v,p} &\leq ||f-g||_{K,p}\cdot||v|| + ||f-g||_p\cdot||v||_{X-K} \\
&< \varepsilon/2 + (M + ||f||_p)\cdot||v||_{X-K} \\
&< \varepsilon/2 + \varepsilon/2 = \varepsilon.
\end{aligned}
$$

Hence f is in the ω_V-closure of S. Q.E.D.

PROPOSITION 5.31 *Let U be a directed set of weights on X and Φ be a filter over $FU_\infty(X,E)$. A sufficient condition for Φ to be convergent is that Φ be a Cauchy filter which converges pointwise. If $U > 0$, this condition is also necessary.*

PROOF When $U > 0$ the condition is obviously necessary, since the topology of pointwise convergence is then weaker than ω_U.

Conversely, let Φ be a Cauchy filter over the space $FU_\infty(X,E)$ which converges pointwise to a function f_0. Let N be a closed neighborhood of the origin in E and $u \in U$. There exists a set $H \in \Phi$ such that $u(x)(f(x) - g(x)) \in N$ for all f and g in H and $x \in X$. For any point $x \in X$, we have then $u(x)(f_0(x)-g(x)) \in N$ for all $g \in H$, since Φ converges pointwise to f_0 and N is closed. Therefore $f_0 \in FU_\infty(X,E)$ and it is the limit of Φ in the space $FU_\infty(X,E)$.

THEOREM 5.32 *If E is complete and $U > 0$, then $FU_\infty(X,E)$ is complete.*

PROOF Let Φ be a Cauchy filter over the space $FU_\infty(X,E)$. By Proposition 5.31, it suffices to prove that Φ converges pointwise. Given $x \in X$, let $u \in U$ be such that $u(x) > 0$. Given $\varepsilon > 0$

and p a continuous seminorm on E, there exists $H \in \Phi$ such that $u(t)p(f(t) - g(t)) < \varepsilon u(x)$, for all $f,g \in H$ and $t \in X$. In particular, $p(f(x) - g(x)) < \varepsilon$ for all $f,g \in H$. Therefore $\Phi(x)$ is a Cauchy filter and thus converges in the space E, since we have assumed E to be complete.

COROLLARY 5.33 *If* E *is complete then the spaces* B(X,E) *and* $B_0(X,E)$ *both equipped with the topology of uniform convergence are complete.*

When E is complete and $V > 0$ on X, the space $FV_\infty(X,E)$ is complete and the Nachbin space $CV_\infty(X,E)$ is therefore complete if and only if it is closed in $FV_\infty(X,E)$.

PROPOSITION 5.34 *If* U *and* V *are two directed sets of weights on* X *with* $U \leq V$, *then* $CU_\infty(X,E)$ *closed in* $FU_\infty(X,E)$ *implies that* $CV_\infty(X,E)$ *is closed in* $FV_\infty(X,E)$.

PROOF Let $f \in FV_\infty(X,E)$ belong to the closure of $CV_\infty(X,E)$ in $FV_\infty(X,E)$. Then f is the limit in $FV_\infty(X,E)$ of a filter Φ in $CV_\infty(X,E)$. From Proposition 5.28, f is also the limit of Φ in the weaker topology ω_U. Since $CV_\infty(X,E) \subset CU_\infty(X,E)$ and $CU_\infty(X,E)$ is closed in $FU_\infty(X,E)$, it follows that $f \in CU_\infty(X,E)$, i.e., $f \in C(X,E)$. Hence, $f \in CV_\infty(X,E)$, and $CV_\infty(X,E)$ is closed in $FV_\infty(X,E)$.

THEOREM 5.35 *Suppose that* E *is complete, and* U *and* V *are two directed sets of weights on* X *with* $U \leq V$. *Then, if* $V > 0$ *on* X *and* $CU_\infty(X,E)$ *is closed in* $FU_\infty(X,E)$, *the Nachbin space* $CV_\infty(X,E)$ *is complete.*

PROOF By Theorem 5.32 the space $FV_\infty(X,E)$ is complete, since $V > 0$ on X. By Proposition 5.34 $CV_\infty(X,E)$ is closed in $FV_\infty(X,E)$, since $CU_\infty(X,E)$ is closed in $FU_\infty(X,E)$ by the hypothesis made.

COROLLARY 5.36 *(Buck) If* X *is a locally compact Hausdorff space and* E *is complete,* $C_b(X,E)$ *equipped with the strict topology is complete.*

PROOF The strict topology on $C_b(X,E)$ is obtained by taking $V = C_0^+(X)$. Now $V > 0$ on X, and if we take $U = K^+(X)$, then $CU_\infty(X,E)$ is $C(X,E)$, which is closed in $F(X,E)$ equipped with the compact open topology (Bourbaki, Topologie Générale, Chapitre 10, § 1). Obviously $K^+(X) \subset C_o^+(X)$.

THOEREM 5.37 *Suppose that* E *is complete and* U *and* V *are two directed sets of weights on* X *with* $U \leq V$. *If* $V > 0$ *on* X *and the Nachbin space* $CU_\infty(X,E)$ *is quasi-complete, then the Nachbin space* $CV_\infty(X,E)$ *is quasi-complete.*

PROOF Let $A \subset CV_\infty(X,E)$ be a closed and bounded subset: Let Φ be a Cauchy filter in A. By Theorem 5.32, the space $FV_\infty(X,E)$ is complete. Hence, there exists $f \in FV_\infty(X,E)$, such that f is the limit of Φ in ω_V. By Proposition 5.28, f is also the limit of Φ in the weaker topology ω_U. Since A is ω_U-bounded, f belongs to $CU_\infty(X,E)$, because $CU_\infty(X,E)$ is quasi-complete by hypothesis. Hence $f \in C(X,E)$, i.e., $f \in CV_\infty(X,E)$. The set A is closed and therefore $f \in A$, i.e., A is complete.

§ 5 DUAL SPACES OF NACHBIN SPACES

Throughout this paragraph X will be a locally compact Hausdorff space. In this case, for any directed set of weights V on X, the space K(X,E) is densely contained in the Nachbin space $CV_\infty(X,E)$. In fact, even $K(X) \otimes E$ is densely contained in $CV_\infty(X,E)$.

Let E'_w denote the topological dual of E endowed with the weak *-topology $\sigma(E',E)$. An E'_w-valued bounded Radon measure u on X is by definition a continuous linear mapping u from K(X) into E'_w, when K(X) is endowed with the topology of uniform convergence on X, given by the sup-norm $||\phi||_\infty = \sup\{|\phi(x)|;\ x \in X\}$ for $\phi \in K(X)$. Any continuous linear functional L on C(X,E) defines an E'_w-valued bounded Radon measure u on X, if we define $u(\phi)$ for each $\phi \in K(X)$ by the formula

$$\langle y,u(\phi)\rangle = L(\phi \otimes y) \tag{1}$$

for all $y \in E$. Conversely, following A. Grothendieck, Produits tensoriels topologiques et espaces nucléaires , Memoirs Amer. Math. Soc. Nº 16 (1955), an E'_w-valued bounded Randon measure u on X is called *integral* if the linear form L defined over $K(X) \otimes E$ by

$$L(\Sigma\ \phi_i \otimes y_i) = \Sigma <y_i, u(\phi_i)> \tag{2}$$

is continuous in the topology induced by $C_o(X,E)$, in which case it can be uniquely continuously extended to $C_o(X,E)$, since $K(X) \otimes E$ is dense in $C_o(X,E)$. In order to characterize the dual of $C_o(X,E)$ as a set of E'_w-valued bounded Radon measures on X, we have fo find necessary and sufficient conditions for such measures to be integral. If $L \in C_o(X,E)'$ let us return to the E'_w-valued bounded Radon measure u definedy by (1). The transpose u' of u is then a linear map from E into $M_b(X)$, the space of all bounded Radon measures on X. Hence, for every $y \in E$ there corresponds a unique regular Borel measure μ_y such that $\mu_y(B) = <u'(y), \chi_B>$, for all Borel subsets B of X. Since L is continuous there exists a continuous seminorm p on E and a constant $k > 0$ such that $|L(f)| \leq k||f||_p$ for all $f \in C_o(X,E)$. Hence

$$|<y, u(\phi)>| = |L(\phi \otimes y)| \leq k\ p(y)\ ||\phi||_\infty .$$

Therefore, the bounded Radon measure u'(y) has norm $||u'(y)|| \leq k\ p(y)$, and the corresponding Borel measure μ_y is such that

$$|\mu_y(B)| \leq ||\mu_y|| \leq k\ p(y).$$

This shows that, for a fixed Borel subset $B \subset X$, the map $y \to \mu_y(B)$ belongs to E'. Call this map $\mu(B)$. The set function $B \to \mu(B)$, defined on the σ-ring of all Borel subsets of X and with values on E' is then countably additive. Indeed, if $\{B_n\}$ is a countable family of disjoint Borel subsets of X and B denotes its union, then for an arbitrary $y \in E$ we have

$$<y, \mu(B)> = \mu_y(B) = \Sigma_{n=1}^{\infty} \mu_y(B_n)$$

$$= \Sigma_{n=1}^{\infty} <y, \mu(B_n)>.$$

This shows that $\mu(B) = \Sigma_{n=1}^{\infty} \mu(B_n)$ in the sense of E'_w. For any finite families $\{B_i\}_{i \in I}$ of disjoint Borel subsets of X, whose union is X, and $\{y_i\}_{i \in I}$ of elements of E with $p(y_i) \leq 1$ for each $i \in I$, we have

$$|\sum_{i \in I} <y_i, \mu(B_i)>| \leq k. \tag{3}$$

An E'_w-valued bounded Radon measure u on X such that the corresponding set function μ satisfies (3) for some continuous seminorm p on E and some constant $k > 0$ is said to have *finite p-semivariation*.

On the other hand, following J. Dieudonné, Sur le théorème de Lebesgue-Nikodym. V, Canad. J. Math. 3 (1951), 129-139, an E'_w-valued bounded Radon measure on X is said to be *p-dominated* if there is a positive bounded Radon measure μ on X such that

$$|\langle y, u(\phi)\rangle| \leq \mu(|\phi|)p(y) \tag{4}$$

for all $y \in E$ and $\phi \in K(X)$.

The arguments contained in I. Singer, Sur les applications linéaires intégrales des espaces de fonctions continues. I, Rev. Math. Pures Appl. 4 (1959), 391-401, and N. P. Các, Linear transformations on some functional spaces, Proc. London Math. Soc.(3) 16 (1966), 705-736, for Banach spaces E can be extended to prove the following.

THEOREM 5.38 *Let* u *be an* E'_w*-valued bounded Radon measure on* X. *Then the following are equivalent:*

(a) u *is integral,*

(b) u *is* p*-dominated, for some continuous seminorm* p *on* E,

(c) u *has finite* p*-semivariation, for some continuous seminorm* p *on* E.

We denote by $M_b(X,E')$ the set of all E'_w-valued bounded Radon measures on X which satisfy (a) or (b) or (c).

COROLLARY 5.39 *The correspondence* $L \leftrightarrow u$ *set up by formulas* (1) *and* (2) *is a vector isomorphism between* $C_o(X,E)'$ *and* $M_b(X,E')$.

The arguments in Wells [67] show that

THEOREM 5.40 *The correspondence* $L \leftrightarrow u$ *set up by formulas* (1) *and* (2) *is a vector isomorphism between* $C_b(X,E),\beta)'$ *and* $M_b(X,E')$.

We apply the above results to characterize the dual of the Nachbin space $CV_\infty(X,E)$ for V in a certain interval of directed sets of weights, following the same path as W.H.Summers, A representation theorem for biequicontinuous completed tensor products of weighted spaces, Trans. Math. Soc. 146 (1969), 121-131.

THEOREM 5.41 *Let* V *be a directed set of weights on* X *with* $C_o^+(X) \leq V \subset B(X)$. *Then the correspondence* $L \leftrightarrow u$ *set up by formulas* (1) *and* (2) *is a vector isomorphism between* $CV_\infty(X,E)'$ *and* $M_b(X,E')$.

PROOF Let $L \in CV_\infty(X,E)'$. Since $V \subset B(X)$, it follows from Proposition 5.28 that $C_o(X,E) \subset CV_\infty(X,E)$ and the topology induced by ω_V is weaker than the uniform topology σ. Hence the restriction of L to $C_o(X,E)$, say M, belongs to $C_o(X,E)'$. According to Corollary 5.39, if we define $u(\phi)$ for each $\phi \in K(X)$ by

$$\langle y,u(\phi)\rangle = M(\phi \otimes y) = L(\phi \otimes y) \tag{1}$$

for all $y \in E$, then $u \in M_b(X,E')$.

Conversely, let $u \in M_b(X,E')$. By Theorem 5.40, if we define L over $K(X) \otimes E$ by

$$L(\Sigma\, \phi_i \otimes y_i) = \Sigma \langle y_i,u(\phi_i)\rangle \tag{2}$$

then L can be extended uniquely to a β-continuous linear functional over $C_b(X,E)$. Since $C_o^+(X) \leq V$, it follows from Proposition 5.28 that $CV_\infty(X,E) \subset C_b(X,E)$ and that the topology induced by β is weaker than ω_V. Hence L can be extended uniquely to an ω_V-continuous linear functional over $CV_\infty(X,E)$. The correspondence set up by (1) and (2) is obviously one-to-one and linear. This ends the proof.

We turn now to the general case of arbitrary Nachbin spaces. Consider E_w'-valued Radon measures u on X, i.e., continuous linear mappings u from K(X) into E_w', when K(X) has its usual inductive limit topology. For every $x \in E = (E_w')'$ the mapping $x \circ u$ is a numerical Radon measure defined by

$$\langle \phi,\, x \circ u\rangle = \langle x,u(\phi)\rangle$$

for all $\phi \in K(X)$. A complex or extended real-valued function f is said to be integrable for u if for every $x \in E$ it is integrable for $x \circ u$, in which case u(f) is that element of $E^* = ((E_w')')^*$ for which

$$\langle x,u(f)\rangle = \int_X f \, d(x \circ u)$$

for all $x \in E$.

Similarly, we say that a function f is locally integrable for u if for every $x \in E$ it is locally integrable for $x \circ u$.

THEOREM 5.42 *Let* $CV_\infty(X;E)$ *be a Nachbin space. Then* $VM_b(X,E')$ *is a vector space and there is a linear isomorphism between* $CV_\infty(X,E)'$ *and* $VM_b(X,E')$.

PROOF Let $v \in V$ and $u \in M_b(X,E')$. Since v is locally integrable for u, and therefore $v\, u(\phi) = u(v\, \phi) \in E^*$ for all $\phi \in K(X)$, let us define a linear functional on $K(X) \otimes E$ by

$$L(\Sigma\, \phi_i \otimes x_i) = \Sigma \langle x_i, v\, u(\phi_i)\rangle.$$

There exists a continuous seminorm p on E and a positive bounded Radon measure μ on X such that

$$|\langle x,u(\phi)\rangle| \leq \mu(|\phi|).p(x) \tag{4}$$

for all $x \in E$ and $\phi \in K(X)$. Let $L_1(X,\mu)$ be the space of all μ-integrable functions, with its usual L_1-seminorm and $L_1(X,\mu,E_p)$ the space of all E_p-valued functions which are μ-integrable with the seminorm

$$||f||_1 = \mu(p \circ f),$$

where E_p denotes E endowed with the seminorm p only.

By (4), we can extend u to a continuous linear mapping t from $L_1(X,\mu)$ into $E_p' \subset E'$. Define a linear functional T on $L_1(X,\mu) \otimes E$ by

$$T(\Sigma\, \psi_i \otimes x_i) = \Sigma\, \langle x_i, t(\psi_i)\rangle$$

For the step functions $f = \Sigma\, \psi_i \otimes x_i$ where the ψ_i's are characteristic functions of pairwise disjoint Borel subsets B_i of X, we have

$$\begin{aligned} |T(f)| &= |\Sigma\, \langle x_i, t(\psi_i)\rangle| \leq \Sigma |\langle x_i, t(\psi_i)\rangle| \\ &\leq \Sigma\, \mu(\psi_i)p(x_i) = \mu(\Sigma\, \psi_i p(x_i)) \\ &= \mu(p \circ f) = ||f||_1. \end{aligned}$$

Hence T is bounded on a dense subspace of $L_1(X,\mu,E_p)$ and we can extend it continuously to $L_1(X,\mu,E_p)$ and still have

$$|T(f)| \leq ||f||_1$$

for all $f \in L_1(X,\mu,E_p)$. In particular, if

$$f = v(\Sigma\ \phi_i \otimes x_i) \in v(K(X) \otimes E)$$

then

$$|L(\Sigma\ \phi_i \otimes x_i)| = |T(f)| \leq ||f||_1 = \mu(v\ p \circ (\Sigma\ \phi_i \otimes x_i)) \leq$$
$$\leq ||\mu|| \ . \ ||v\ p \circ (\Sigma\ \phi_i \otimes x_i)||_\infty = ||\mu||.||\Sigma\ \phi_i \otimes x_i||_{v,p}.$$

Therefore L is continuous on $K(X) \otimes E$ with the topology induced by $CV_\infty(X,E)$ and can be uniquely continuously extended to $CV_\infty(X,E)$.

Conversely, let L be a continuous linear functional on $CV_\infty(X,E)$. Define t on K(X) by

$$<x,t(\phi)> = L(\phi \otimes x)$$

for all $x \in E$. Obviously, $t(\phi) \in E^*$. Since L is continuous there exists $v \in V$ and a continuous seminorm p on E such that

$$|L(f)| \leq ||f||_{v,p}$$

for all $f \in CV_\infty(X,E)$. Therefore, if $f = \phi \otimes x$ then

$$|<x,t(\phi)>| = |L(\phi \otimes x)| \leq ||\phi \otimes x||_{v,p} \leq ||v\ \phi||_\infty . p(x)$$

This proves that $t(\phi) \in E'$. On the other hand, if $K \subset X$ is a compact subset and $\phi \in K(X)$ has support contained in K, then $||v\ \phi||_\infty \leq ||v||_K . ||\phi||_\infty$, which shows that t is an E'_w-valued Radon measure on X. Let w be the positive extended real-valued function 1/v. Then w is lower semicontinuous. Let $x \in E$, and $\mu_x = x \circ t$ the corresponding numerical Radon measure on X. Let $\mu_x = \mu_1 - \mu_2 + i(\mu_3 - \mu_4)$ be the minimal decomposition of μ_x with $\mu_i \geq 0$ $(i = 1,2,3,4)$. Notice that for any $\psi \in K(X)$, with $0 \leq \psi \leq w$, we have $||v\ \psi||_\infty \leq 1$, and $||\psi \otimes x||_{v,p} \leq p(x)$. Since

$$\mu_i(\phi) \leq \sup\ \{|(x \circ t)(\psi)|;\ 0 \leq \psi \leq \phi,\ \psi \in K(X)\}$$

it follows that $\mu_i(\phi) \leq p(x)$, for any $0 \leq \phi \leq w$, as $\psi \leq \phi$ implies then $\psi \leq w$, and therefore $|(x \circ t)(\psi)| \leq ||\psi \otimes x||_{v,p} \leq p(x)$. Hence

$$\mu_i^*(w) = \sup \{\mu_i(\phi);\ 0 \le \phi \le w,\ \phi \in K(X)\} \le p(x),$$

and $w \in L_1(X,\mu_i)$, $i = 1,2,3,4$. Thus $w \in L_1(X, x\circ t)$ for all $x \in E$, i.e., w is integrable with respect to t. Define $u(\phi)$ for each $\phi \in K(X)$ by

$$\langle x, u(\phi)\rangle = \langle w\,\phi,\ x \circ t\rangle$$

for all $x \in E$. Obviously $u(\phi) \in E^*$. Since

$$|\langle w\,\phi,\ x \circ t\rangle| = \left|\int_X w\,\phi\, d(x \circ t)\right|$$
$$\le ||\phi||_\infty . \sqrt{8}\ p(x),$$

it follows that $u(\phi) \in E'$, and in fact u is an E'_w-valued Radon measure on X. Moreover the inequality

$$|\langle x, u(\phi)\rangle| \le \sqrt{8}\ p(x)\, ||\phi||_\infty$$

implies that the corresponding set function $B \to \mu(B)$ satisfies (3) with $k = \sqrt{8}$, i.e., $u \in M_b(X,E')$.

It follows that

$$\langle \phi,\ x \circ t\rangle = \langle v\,\phi, x \circ u\rangle$$

for all $\phi \in K(X)$, $x \in E$, i.e., $t = v\,u$. Define $\Phi(L) = v\,u$. We have seen that Φ is one-to-one and onto $VM_b(X,E')$, and that $VM_b(X,E')$ is a set of E'_w-valued Radon measures on X. Notice that $VM_b(X,E')$ is closed under multiplication by scalars. Let $v'u'$, $v''u'' \in VM_b(X,E')$. By the first part of our proof, there exist L',L'' in $CV_\infty(X,E)'$ such that $\Phi(L') = v'u'$ and $\Phi(L'') = v''u''$. Let $\Phi(L' + L'') = v\,u$. Then for every $\phi \in K(X)$ and $x \in E$, we have

$$\langle x, v\,u(\phi)\rangle = (L' + L'')(\phi \otimes x) = L'(\phi \otimes x) + L''(\phi \otimes x)$$
$$= \langle x, v'\,u'(\phi)\rangle + \langle x, v''u''(\phi)\rangle$$

i.e., $v\,u = v'u' + v''u'' \in VM_b(X,E')$ and $\Phi(L' + L'') = \Phi(L') + \Phi(L'')$. The proof that $\Phi(\lambda L) = \lambda\Phi(L)$ is trivial.

This ends the proof of Theorem 5.42.

Let V be a directed set of weights on X such that each $v \in V$ is continuous, i.e., $V \subset C^+(X) = \{f \in C(X;\mathbb{R});\ f \ge 0\}$.

Let $u \in M_b(X;E)$. There is a continuous seminorm p

on E, and a positive and bounded Radon measure μ on X such that

$$|\langle y, u(\phi)\rangle| \leqq \mu(|\phi|)p(y) \tag{1}$$

for all $y \in E$ and $\phi \in K(X)$. Since $u \in M_b(X, E')$, then

$$U(\Sigma \phi_i \otimes y_i) = \Sigma \langle y_i, u(\phi_i)\rangle \tag{2}$$

defined on $K(X) \otimes E$ can be extended to an element of $C_o(X, E)'$ and then

$$|\mu|_p(\phi) = \sup\{|U(f)| ; f \in K(X,E), p(f) \leqq \phi\}$$

defined on $K^+(X)$ can be extended to be a positive and bounded Radon measure on X, which will be the least $\mu \in M_b^+(X)$ satisfying (1). By definition,

$$\|u\|_p = |\mu|_p(1) = \int_X d|\mu|_p .$$

Since by our hypothesis any $v \in V$ is continuous, the operator $T_v(f) = vf$ maps $CV_\infty(X,E)$ into $C_o(X,E)$. Let $D_p = \{g \in C_o(X, E); \|g\|_p \leqq 1\}$, if p is a continuous seminorm on E. Then $D_{v,p} = T_v^{-1}(D_p)$. Let T_v^* denote the transpose map of T_v.

PROPOSITION 5.43: *Let* V *be a directed set of weights on* X *with* $V \subset C^+(X)$. *Then*

$$D_{v,p}^o = T_v^*(D_p^o).$$

PROOF: The operator T_v is a continuous linear map from $CV_\infty(V,E)$ into $C_o(X,E)$, and therefore continuous in the weak topologies. Hence T_v^* is continuous in the corresponding weak *-topologies. Now D_p^o is weak * - compact by Alaoglu's Theorem, and therefore $T_v^*(D_p^o)$ is also weak * - compact. Since $T_v^*(D_p^o)$ is absolutely convex, it follows from the Bi - polar Theorem that

$T_v^*(D_p^o) = [T_v^*(D_p^o)]^{oo}$. On the other hand, $T_v^{-1}(D_p^{oo}) = [T_v^*(D_p^o)]^o$ and $D_p^{oo} = D_p$ (the last inequality follows again by the Bipolar Theorem). Hence

$$D_{v,p}^o = [T_v^{-1}(D_p)]^o = [T_v^{-1}(D_p^{oo})]^o =$$

$$= [T_v^*(D_p^o)]^{oo} = T_v^*(D_p^o).$$

COROLLARY 5.44: *If* $V \subset C^+(X)$, *then* $H \subset CV_\infty(X,E)'$ *is equicontinuous if, and only if, there exists* $v \in V$ *and* p *a continuous seminorm on* E *such that* $H \subset T_v^*(D_p^o)$.

COROLLARY 5.45: *If* $V \subset C^+(X)$, *then*

$$D_{v,p}^o = v \cdot \{u \in M_b(X,E'); \|u\|_p \leqq 1\}.$$

PROOF: If p is a continuous seminorm on E, then $D_p^o = \{u \in M_b(X,E'); \|u\|_p \leqq 1\}$ follows easily from Corollary 5.39. By Proposition 5.43 above, $D_{v,p}^o = T_v^*(D_p^o)$; while it is clear that

$$T_v^*(\{u \in M_b(X,E'); \|u\|_p \leqq 1\}) = v \cdot \{u \in M_b(X,E'); \|u\|_p \leqq 1\}.$$

THEOREM 5.46: *Let* W *be a vector subspace of* $CV_\infty(X,E)$ *and let* $L \neq 0$ *be an extreme point of* $W^\perp \cap D_{v,p}^o$. *If for* $g \in C(X)$ *the restriction of* g *to the support of* L *is bounded and real-valued, while* $L(gw) = 0$ *for every* $w \in W$, *then* g *is constant on the support of* L.

PROOF: Let $L \neq 0$ be an extreme point of $W^\perp \cap D_{v,p}^o$. By Corollary 5.45 above, $L = vu$, where $u \in M_b(X,E')$ is such that $\|u\|_p \leqq 1$. Since L is extreme, it follows that $\|u\|_p = 1$. We may assume without loss of generality that u and vu have the same support and that $0 \leqq g \leqq 1$ on the support of u. Let $e = gu$. Since $|g(x)\phi(x)| \leqq |\phi(x)|$ for all $x \in X$ and $\phi \in K(X)$, it follows that $e \in M_b(X,E')$. Moreover, $|e|_p = g|\mu|_p$ and since u and $|\mu|_p$ have the same support:

$$\|e\|_p = |e|_p(1) = \int g d|\mu|_p = \int g X_F d|\mu|_p \leqq \|\mu\|_p = 1$$

where F denotes the support of $|\mu|_p$ and X_F its characteristic function. If $\|e\|_p = 0$, then

$$\int g X_F d|\mu|_p = 0$$

and $g = 0$ on the support of $|\mu|_p$. Similarly if $\|e\|_p = 1$, then $g = 1$ on the support of $|\mu|_p$. So we may assume $0 < \|e\|_p < 1$. Let $\sigma = \|e\|_p^{-1} e$ and $\tau = (1 - \|e\|_p)^{-1}(u - e)$. Let $S = v\sigma$ and $T = v\tau$. Cleary S belongs to $W^\perp \cap D^o_{v,p}$, while $T = (1 - \|e\|_p)^{-1}(L - R)$, where $R = ve$, implies $T \in W^\perp$. On the other hand, we have

$$\|\tau\|_p = (1 - \|e\|_p)^{-1} \int (1 - g) d|\mu|_p$$

$$= (1 - \|e\|_p)^{-1}(\|\mu\|_p - \|e\|_p)$$

$$= 1 .$$

Hence $\tau \in D^o_p$ and consequently $T \in D^o_{v,p}$. But $vu = (1-\|e\|_p)v\tau + \|e\|_p v\sigma$, i.e. $L = (1 - \|e\|_p)T + \|e\|_p S$. Since L is an extreme point, $L = T = S$. From this it follows that $\|e\|_p vu = ve = gvu$, and therefore g is constant on the support of vu, which is by definition the support of L.

THEOREM 5.47: *Let* V *be a directed set of weights on* X *with* $V \subset C^+(X)$. *Let* A *be a subalgebra of* C(X) *such that every* $g \in A$ *is bounded on the support of every* $v \in V$. *Let* W *be a vector subspace of* $CV_\infty(X,E)$ *which is an* A-*module. Then* $f \in CV_\infty(X,E)$ *is in the closure of* W *if and only if* $f|K$ *is in the closure of* $W|K$ *in* $CV_\infty(K,E)$ *for each* $K \in \mathcal{K}_A$.

PROOF: The proof is analogous to that of Theorem 3.2, Prolla [51]. Notice that if $L = vu$, where u is chosen so that the support of L is equal to the support of u, then L defines a continuous linear functional over $CV_{\infty}(K,E)$ for any closed subset K of X which contains the support of u.

APPENDIX

FUNDAMENTAL WEIGHTS

The proof of Theorem 5.19 relies on showing that a certain function γ on $\mathbb{R}$ is a fundamental weight (see Definition 5.5) in the sense of S. Bernstein. This is done by an appeal to the Denjoy-Carleman Theorem on quasi-analytic classes. This is the reason for the name "quasi-analytic criterion" given to Theorem 5.19. Conversely, weighted approximation techniques can be used to solve the problem of characterizing quasi-analytic classes of functions. In this appendix we present a very simple proof, due to G. Zapata, of Mergelyan's theorem which characterizes the fundamental weights on the real line. This result was then used by Zapata to show that Hadamard's problem on the characterization of quasi-analytic classes of functions is equivalent to S. Bernstein's problem on the characterization of fundamental weights. Using this equivalence, one gets a solution of Hadamard's problem by the sole means of weighted approximation theory. As pointed out by Zapata, this seems to be an interesting approach to the generalized quasi-analytic problem of Mandelbrojt. (See S. Mandelbrojt, "Séries adhérentes, régularisation des suites, applications", Paris, Gauthier-Villars,1952).

§ 1 MERGELYAN'S THEOREM

Let ω be a *weight* on $\mathbb{R}$, i.e. an upper semicontinuous positive function defined on $\mathbb{R}$. Recall from Definition 5.1 that the Nachbin space $C\omega_\infty(\mathbb{R})$ is the vector space of all continuous and complex valued functions f on $\mathbb{R}$ such that ωf vanishes at infinity. The topology of $C\omega_\infty(\mathbb{R})$ is determined by the seminorm

$$f \to \sup \{\omega(x)|f(x)|;\ x \in \mathbb{R}\} = ||f||_\omega$$

A weight ω on $\mathbb{R}$ is said to be *rapidly decreasing at infinity* when the set $\mathcal{P}(\mathbb{R})$ of complex valued polynomials on

$\mathbb{R}$ is contained in $C\omega_\infty(\mathbb{R})$. (See Definition 5.5). The weight ω is called *fundamental* if $(\mathbb{R})$ is densely contained in $C\omega_\infty(\mathbb{R})$, i.e. for every $\varepsilon > 0$ and every $f \in C\omega_\infty(\mathbb{R})$ there exists a polynomial $p \in (\mathbb{R})$ such that

$$\omega(x)\,|f(x) - p(x)| < \varepsilon$$

for all $x \in \mathbb{R}$.

The problem of determining necessary and sufficient conditions for ω to be fundamental was first stated by S. Bernstein in 1924. (See S. Bernstein, "Le problème de l'approximation des fonctions continues sur tout l'axe réel et l'une de ses applications", Bulletin de la Société Mathématique de France, tome 52 (1924), 399-410).

REMARK 1 For weights ω and ν on $\mathbb{R}$, if $\omega \leq \nu$ and ν is a fundamental weight, then ω is a fundamental weight too. (See Proposition 1, § 24, Nachbin [43]). Also, if the weight ω is *bounded* on $\mathbb{R}$, then

$$C_0(\mathbb{R}) \subset C\omega_\infty(\mathbb{R}).$$

In this case $C_0(\mathbb{R})$ is dense in $C\omega_\infty(\mathbb{R})$, because $K(\mathbb{R}) \subset C_0(\mathbb{R})$ and $\mathbb{R}$ is locally compact. (See Proposition 2, § 22, Nachbin [43]).

Notice also that the inclusion map is continuous, when $C_0(\mathbb{R})$ carries the uniform topology σ. Indeed,

$$\|f\|_\omega = \|\omega f\| = \|\omega\| . \|f\|$$

for all $f \in C_0(\mathbb{R})$, where we have set

$$\|h\| = \sup \{|h(x)|;\ x \in \mathbb{R}\}$$

for all $h: \mathbb{R} \to \mathbb{C}$ which is bounded on $\mathbb{R}$.

REMARK 2 Let $D = \{z \in \mathbb{C};\ \mathrm{Im}(z) \neq 0\}$. For $z \in D$, let

$$g_z(t) = \frac{1}{t-z}$$

for all $t \in \mathbb{R}$. Let $A = \mathbb{C}[g_z, \bar{g}_z]$ be the complex algebra generated by the functions g_z and $\bar{g}_z$. Notice that $A \subset C_0(\mathbb{R})$. We claim that A is dense in $C_0(\mathbb{R})$. Indeed, the algebra A is separating, everywhere different from zero on $\mathbb{R}$, and is self-adjoint. Since $C_o(\mathbb{R}) \subset C_b(\mathbb{R})$, we may apply Theorem 5.20 (or Corollary 1.9 to a one-point compactification of $\mathbb{R}$) to conclude that A is dense in $C_o(\mathbb{R})$. As a corollary, if ω is bounded, then A is dense in $C\omega_\infty(\mathbb{R})$, by Remark 1.

REMARK 3: Let $a, b \in \mathbb{R}$, $a \neq 0$, and let

$$\sigma(t) = at + b$$

for all $t \in \mathbb{R}$. For any $f : \mathbb{R} \quad \mathbb{C}$, let

$$T(f) = f \circ \sigma \quad .$$

Then $T(K(\mathbb{R})) = K(\mathbb{R})$ and

$$T(\mathcal{P}(\mathbb{R})) = \mathcal{P}(\mathbb{R}) .$$

Also, the restriction of T to $C\omega_\infty(\mathbb{R})$ in an isometry of $C\omega_\infty(\mathbb{R})$ onto $C(\omega \circ \sigma)_\infty(\mathbb{R})$. Hence, the weight ω is fundamental if, and only if $\omega \circ \sigma$ is fundamental.

DEFINITION 1. *Let* ω *be a weight on* $\mathbb{R}$. *Let* $\mathcal{P}_\omega = \{P \in (\mathbb{R}); \|\omega P\| \leq 1\}$. *For every* $z \in \mathbb{C}$, *let*

$$M_\omega(z) = \sup\{|P(z)| \ ; \ P \in \mathcal{P}_\omega\}.$$

DEFINITION 2. *Let* ω *be a weight on* $\mathbb{R}$, *For every* $t \in \mathbb{R}$, *let*

$$\omega^*(t) = \frac{\omega(t)}{1+|t|}$$

In 1954, using the function M_{ω^*} introduced by himself, Mergelyan was able to give a necessary and sufficient condition for ω to be fundamental, thus solving Bernstein's problem for the real line. His proof, as well as a survey of earlier results on Bernstein's problem, is available in English in a translation of R. P. Boas, Jr. (See S. N. Mergelyan, "Weighted approximation by polynomials", American Methematical Society Translations, Series 2, volume 10 (1958), pp. 59 - 106).

THEOREM 1. (MERGELYAN): *A weight* ω *on* $\mathbb{R}$ *is fundamental if, and only if,* $M_{\omega^*}(z) = +\infty$ *, for some* $z \in D$.

PROOF: (a) *Necessity*. Let ω be a fundamental weight on $\mathbb{R}$. Let $z \in D$, and let $\varepsilon > 0$ be given. Since any fundamental weight on $\mathbb{R}$ is bounded, $C_0(\mathbb{R}) \subset C\omega_\infty(\mathbb{R})$, by Remark 1. Therefore $g_z \in C\omega_\infty(\mathbb{R})$. Let $P \in \mathcal{P}(\mathbb{R})$ be such that

$$(1) \qquad \| g_z - P \|_\omega < \varepsilon .$$

Let $K = \inf \{ (1 + |x|) \cdot | g_z(x) | \; ; \; x \in \mathbb{R} \}$. Then the polynomial defined by

$$Q(x) = \frac{K}{\varepsilon} (1 - (x - z)P(x))$$

for all $x \in \mathbb{R}$, is such that $\|\omega^* Q\| \leq 1$. Indeed, for any $x \in \mathbb{R}$, we have

$$\omega^*(x)|Q(x)| = \frac{\omega(x)}{1+|x|} \cdot |1 - (x - z)P(x)| \cdot \frac{K}{\varepsilon}$$

$$\leq \frac{\omega(x) \, | g_z(x) - P(x)|}{(1 + |x|)|g_z(x)|} \cdot \frac{K}{\varepsilon} \leq 1 .$$

Therefore $M_{\omega^*}(z) \geq |Q(z)| = K/\varepsilon$. Since $\varepsilon > 0$ was arbitrary, $M_{\omega^*}(z) = +\infty$.

(b) *Sufficiency*.

We will present Zapata's proof (see Zapata [68]), which rests on the following two lemmas.

LEMMA 1. *Let* ν *be a weight on* $\mathbb{R}$ *such that* $M_\nu(z) = +\infty$ *for some* $z \in D$. *Then* ν *is rapidly decreasing at infinity*.

PROOF: Let W be the vector space of all $P \in \mathcal{P}(\mathbb{R})$ such that νP is bounded. For any such P, put $\|P\|_\nu = \|\nu P\|$. Then $\|\cdot\|_\nu$ is a semi-norm on W. Assume that $\|\cdot\|_\nu$ is not a norm. Then there exists $P \in \mathcal{P}(\mathbb{R})$, $P \neq 0$, such that $\nu P = 0$. This implies that ν has a finite set as its support, whence $W = \mathcal{P}(\mathbb{R})$. Assume now that $\|\cdot\|_\nu$ is a norm. Assume that W has finite dimension, say m. We have $m \geq 1$ since $m = 0$ would imply $M_\nu(z) = 0$. Notice that $P \in W$ and $n = \text{degree}\,(P)$ imply $t^n \in W$, whence the set $\{1,\ldots,t^{m-1}\}$ is a basis for W and the mapping $\psi_i : W \to \mathbb{C}$ given by $\psi_i(a_o + \ldots + a_{m-1}t^{m-1}) = a_i$ is continuous for $i=0,\ldots,m-1$. Also, the set $\mathcal{P}_\nu$ is closed and bounded, hence $\mathcal{P}_\nu$ is compact by our assumption. From the compactness of $\psi_i(\mathcal{P}_\nu)$, there exists a positive constant C such that for all $P \in \mathcal{P}_\nu$, $P = a_o + \ldots + a_{m-1}\, t^{m-1}$, we have $|a_i| \leq C$ for all $i=0,\ldots,m-1$. Thus $|P(z)| \leq C \sum_{i=0}^{m-1} |z|^i$ for all $P \in \mathcal{P}_\nu$; that is $M_\nu(z) < +\infty$, contradicting the hypothesis. Hence W has infinite dimension. Since $t^n \in W$; implies $t^m \in W$ for all $0 \leq m \leq n$, it follows that $W = \mathcal{P}(\mathbb{R})$. To finish the proof, notice that ν is rapidly decreasing at infinity if, and only if, νP is bounded on $\mathbb{R}$ for every $P \in \mathcal{P}(\mathbb{R})$. (See Definition 1, § 24, Nachbin [43]).

LEMMA 2: *Let ω be a weight on $\mathbb{R}$, which is rapidly decreasing at infinity. If, for some $z \in D$, $g_z \in \overline{\mathcal{P}(\mathbb{R})}$, then $\mathbb{C}[g_z,\bar{g}_z] \subset \overline{\mathcal{P}(\mathbb{R})}$.*

PROOF: Let $W = \overline{\mathcal{P}(\mathbb{R})}$. Since W is a vector subspace of $C\omega_\infty(\mathbb{R})$, it is enough to prove that $g_z^n(\bar{g}_z)^m \in W$ for all $n,m \in \mathbb{N}$. First, we will prove by induction that

$$(2) \qquad g_z^n\,\mathcal{P}(\mathbb{R}) \subset W \quad \text{for all} \quad n \in \mathbb{N}\,.$$

In fact, for $n = 0$ this is clear. Assume that it is true for some n and let $P \in \mathcal{P}(\mathbb{R})$. Since $Q = g_z(P - P(z)) \in (\mathbb{R})$ and $g_z^{n+1}(P - P(z)) = g_z^n Q$, we have

$$(3) \qquad g_z^{n+1}(P - P(z)) \in W$$

Note that if g is a continuous and bounded complex valued function on $\mathbb{R}$, then the mapping $f \to gf$ from $C\omega_\infty(\mathbb{R})$ into itself is continuous. So

$$(4) \qquad g\bar{S} \subset \overline{gS} \quad \text{for all} \quad S \subset C\omega_\infty(\mathbb{R})\,.$$

Since $g_z^{n+1} \in g_z^n W = g_z^n\,\overline{\mathcal{P}(\mathbb{R})}$, the above remark and the induction hypothesis imply $g_z^{n+1} \in W$. Then, from $g_z^{n+1} P = g_z^{n+1}(P - P(z)) + P(z)g_z^{n+1}$ and (3), we get $g_z^{n+1} P \in W$. Thus (2) is proved. Notice that $h \in W$ implies $\bar{h} \in W$ whence, from (2), $(\bar{g}_z)^m = \overline{g_z^m} \in W$ for all $m \in \mathbb{N}$, and then, from (4), $g_z^n(\bar{g}_z)^m \in g_z^n \mathcal{P}(\mathbb{R}) \subset \overline{g_z^n \mathcal{P}(\mathbb{R})}$ for all $n,m \in \mathbb{N}$. Using (2) and the fact that W is closed, we conclude that $g_z^n(\bar{g}_z)^m \in W$ for all $n,m \in \mathbb{N}$.

Proof of *Sufficiency:* We have $M_{\omega^*}(z) = +\infty$ for some $z \in D$. From Lemma 1, it follows that $(1 + t^2)\mathcal{P}(\mathbb{R}) \subset C\omega_\infty^*(\mathbb{R})$ whence $\mathcal{P}(\mathbb{R}) \subset C\omega_\infty(\mathbb{R})$. Let $Q \in \mathcal{P}_{\omega^*}$ be such that $Q(z) \neq 0$, and put

$$P = \frac{g_z}{Q(z)} (Q - Q(z)).$$

Then $P \in \mathcal{P}(\mathbb{R})$ and $g_z - P = g_z Q / Q(z)$. Hence, $\omega(t)|g_z(t)-P(t)| \leq \leq |Q(z)|^{-1}(1 + |t|)\, |g_z(t)|\, \omega^*(t)\, |Q(t)| \leq$ constant. $|Q(z)|^{-1}$ for all $t \in \mathbb{R}$. Letting $Q(z) \to \infty$, we have that $\|\omega(g_z - P)\| \to 0$. So $g_z \in \overline{\mathcal{P}(\mathbb{R})}$. From Lemma 2, it follows that $\mathbb{C}[g_z, \bar{g}_z] \subset \overline{\mathcal{P}(\mathbb{R})}$. Since ω is bounded, by Remark 2 we have that $\mathbb{C}[g_z, \bar{g}_z]$ is dense in $C\omega_\infty(\mathbb{R})$, and so we conclude that $\mathcal{P}(\mathbb{R})$ is dense in $C\omega_\infty(\mathbb{R})$.

REMARK 4: From the proof of the necessity of the condition in Theorem 1, it follows that $M_{\omega^*}(z) = +\infty$ for every $z \in D$, whenever ω is a fundamental weight.

§ 2. FUNDAMENTAL WEIGHTS AND QUASI-ANALYTIC CLASSES OF FUNCTIONS

In the following, M will denote a sequence (M_n), $n \in \mathbb{N}$, of positive real numbers.

We will denote by C(M) the class of all complex valued indefinitely differentiable functions f on $\mathbb{R}$ such that there exist positive constants C and c (depending on f) for which

$$|f^{(n)}(t)| \leq C\, c^n M_n \quad \text{for all } t \in \mathbb{R},\ n \in \mathbb{N}.$$

The class C(M) is called *quasi-analytic* if the following condition holds: $f \in C(M)$ vanishes identically, if there exists $s \in \mathbb{R}$ such that $f^{(n)}(s) = 0$ for all $n \in \mathbb{N}$.

Hadamard's problem consists in finding necessary and sufficient conditions on a given sequence $M = (M_n)$ in order that the class $C(M)$ be quasi-analytic.

REMARK 5: Let σ be as in Remark 3. Then $C(M) \circ \sigma \subset C(M)$.

We write γ_M for the weight on $\mathbb{R}$ given by

$$\gamma_M(t) = \inf\{ M_n |t|^{-n}, n \in \mathbb{N}\} \quad \text{for all} \quad t \in \mathbb{R}.$$

REMARK 6: We have $\mathcal{P}(\mathbb{R}) \subset C(\gamma_M)_\infty(\mathbb{R})$. Further, either γ_M is an upper-semicontinuous function with compact support (when $\limsup M_n^{1/n} < +\infty$) or it is a continuous function that never vanishes. In the first case, γ_M is fundamental by the Weierstrass polynomial approximation theorem.

THEOREM 2: *The class* $C(M)$ *is quasi-analytic if, and only if,* γ_M *is a fundamental weight.*

PROOF: The necessity of the condition follows from the proof of Lemma 2, §29, Nachbin [43].

Let us prove the sufficiency. Assume that the class $C(M)$ is not quasi-analytic. From Remark 5, there exists $f \in C(M)$ such that

$$(1) \qquad f^{(n)}(0) = 0 \ \text{for all} \ n \in \mathbb{N} \quad \text{and} \quad f|[0,+\infty) \not\equiv 0.$$

Let

$$U = \{z \in C, \ \mathrm{Re}(z) > 0\}, \quad V = \{z \in C, \ \mathrm{Re}(z) > 1\}$$

and put

$$F(z) = \int_0^{+\infty} f(t) e^{-izt}\, dt \quad \text{for all} \quad z \in U.$$

Integrating by parts and using induction, we get, in view of (1),

$$(2) \quad z^n F(z) = (-i)^n \int_0^{+\infty} f^{(n)}(t) e^{-izt}\, dt$$

$$\text{for all} \quad z \in U,\ n \in \mathbb{N}.$$

Since $f \in C(M)$, there exist constants C and c such that

$$(3) \quad |f^{(n)}(t)| \leq C\, c^n\, M_n \quad \text{for all} \quad t \in \mathbb{R},\ n \in \mathbb{N}.$$

From this and (2), it follows that

$$(4) \quad |F(z)| \leq C\, c^n\, M_n |z|^{-n} \quad \text{for all} \quad z \in \overline{V},\ n \in \mathbb{N}.$$

Since F is a holomorphic function not identically zero, we have

$$F(\alpha) \neq 0 \quad \text{for some} \quad \alpha \in V.$$

Let ω be the weight given by

$$\omega(t) = (1 + |t|)|F(1 + it)||1 + it|^{-1} \quad \text{for all} \quad t \in \mathbb{R}.$$

Fix $P \in \mathcal{P}_{\omega^*}$, and let

$$G(z) = P(i - iz)F(z)z^{-1} \quad \text{for all} \quad z \in U.$$

Then G is holomorphic in V and continuous on $\overline{V}$. Letting $t \in \mathbb{R}$, since $|G(1 + it)| = |\omega^*(t)P(t)|$, we have that $|G| \leq 1$ on ∂V because $P \in \mathcal{P}_{\omega^*}$. Also, G is bounded on V from (4). Since $\alpha \in V$, it follows from the maximum modulus theorem that $|G(\alpha)| \leq 1$, whence $|P(i - i\alpha)| \leq |\alpha||F(\alpha)|^{-1}$. Notice that, P being arbitrary, $M_{\omega^*}(i - i\alpha) < +\infty$. Since $i - i\alpha \in D$, Remark 4 implies that ω is not fundamental. Furthermore, (4) implies that

$$\omega(t) \leq \sqrt{2}\, C\, c^n\, M_n |1 + it|^{-n} \leq \sqrt{2}\, C\, M_n \left|\frac{t}{c}\right|^{-n}$$

for all $t \in \mathbb{R}$, $n \in \mathbb{N}$; whence

$$\omega(t) \leq \sqrt{2}\, C\, \gamma_M\left(\frac{t}{c}\right) \quad \text{for all} \quad t \in \mathbb{R}.$$

From this and Remarks 1 and 3, we conclude that γ_M is not fundamental since ω is not fundamental.

COROLLARY 1: *Let* ϕ *be a complex valued* C^∞ *function on* $\mathbb{R}$ *with compact support and not identically zero. Put* $M_n = \| \phi^{(n)} \|$ *for all* $n \in \mathbb{N}$. *Then* γ_M *is not fundamental.*

PROOF: Since $\phi \in C(M)$, the class $C(M)$ is not quasi-analytic and the conclusion follows from Theorem 2.

REMARK 7: The above corollary provides a simple counterexample to localizability (see §31 of Nachbin [43]). Notice also that, in this case, γ_M is a continuous and positive function by Remark 6.

COROLLARY 2: *Let* ω *be a weight on* $\mathbb{R}$ *such that* $\mathcal{P}(\mathbb{R}) \subset C\omega_\infty(\mathbb{R})$ *and* $\mathcal{P}(\mathbb{R})$ *is not dense. Let* $M_n = \| \omega t^n \|$, *for all* $n \in \mathbb{N}$. *Then the class* $C(M)$ *is not quasi-analytic.*

PROOF: Since $\omega \leq \gamma_M$, the conclusion follows from Theorem 2 and Remark 1.

LEMMA 3: *For* M *fixed, let* $\omega(t) = (1 + |t|)\, \gamma_M(t)$ *for all* $t \in \mathbb{R}$. *Then* ω *is fundamental if, and only if,* γ_M *is fundamental.*

PROOF: Since $\gamma_M \leq \omega$, in view of Remark 1 it is enough to prove that ω is fundamental when γ_M is fundamental. In fact, let M' be defined by $M'_n = M_{n+1}$ for all $n \in \mathbb{N}$. for $t \neq 0$, we have

$$\gamma_{M'}(t)\,|t|^{-1} = \inf \{ M_{n+1} |t|^{-(n+1)},\ n \in \mathbb{N} \} \geq \gamma_M(t),$$

and hence

$$|t|\ \gamma_M(t) \leq \gamma_{M'}(t) \quad \text{for all} \quad t \in \mathbb{R}.$$

Also

$$\gamma_M(t) \leq \gamma_M(0) \leq \gamma_M(0)\ \gamma_{M'}(t) / \gamma_{M'}(1)$$

for $|t| \leq 1$. so, there exists a positive constant C such that $\omega \leq C \cdot \gamma_{M'}$. Since γ_M is fundamental, it follows from Theorem 2 that C(M) is quasi-analytic. Then C(M') is a quasi-analytic class, whence $\gamma_{M'}$ is fundamental by Theorem 2. So, from Remark 1, we conclude that ω is fundamental.

Put $T_M(t) = \sup\{|t|^n M_n^{-1},\ n \in \mathbb{N}\}$ for all $t \in \mathbb{R}$.

THEOREM 3: *The class* C(M) *is quasi-analytic if, and only if, there exist a complex number* z *and a sequence of polynomials* (P_n) *such that*

$$\mathrm{Im}(z) \neq 0, \quad |P_n| \leq T_M \ \textit{for all}\ n, \quad \textit{and} \quad P_n(z) \to \infty .$$

PROOF: Let ω be as in Lemma 3. From this and Theorem 2, we have that the class C(M) is quasi-analytic if, and only if, ω is fundamental. Since $\mathcal{P}_{\omega^*} = \{P \in \mathcal{P}(\mathbb{R}),\ |P| \leq T_M\}$, Theorem 3 follows from Theorem 1.

REFERENCES FOR CHAPTER 5.

BUCK [11]

GLICKSBERG [26]

KLEINSTÜCK [35]

MACHADO and PROLLA [39], [40], [41]

NACHBIN [42], [43], [45]

NACHBIN, MACHADO and PROLLA [46]

PROLLA [50], [51]

PROLLA and MACHADO [52]

SUMMERS [64]

TODD [65]

WELLS [67]

ZAPATA [68]

CHAPTER 6

THE SPACE $C_o(X;E)$ WITH THE UNIFORM TOPOLOGY

Let X be a Hausdorff space. If $V = \{v\}$, where $v: X \to \mathbb{R}$ is the constant function 1, then the Nachbin space $CV_\infty(X;E)$ is the space $C_o(X;E)$, for each locally convex space E (see Example 5.2, Chapter 5.) Its topology is the topology of uniform convergence on X.

THEOREM 6.1. *Let $A \subset C_b(X;\mathbb{K})$ be a subalgebra, and let $W \subset C_o(X;E)$ be an A-submodule. Then W is sharply localizable under A in $C_o(X;E)$.*

PROOF. Since $A \subset C_b(X;\mathbb{K})$, we can apply Corollary 5.21, § 3, Chapter 5.

COROLLARY 6.2. *Let A and W be as in theorem 6.1. Assume that A is self-adjoint. Then W is localizable under A in $C_o(X;E)$.*

PROOF. Since A is self-adjoint, then $G(A) = \mathrm{Re}\, A \cup \mathrm{Im}\, A$ is a strong set of generators for A satisfying the hypothesis of theorem 5.20, § 3, Chapter 5. On the other hand, since G(A) consists only of real valued functions, $\rho = 2$ and $P_2 = P_A$. Thus W is localizable under A in $C_o(X;E)$.

COROLLARY 6.3. *Let A and W be as in Corollary 6.2. Assume that A is separating. Then W is dense in $C_o(X;E)$ if, and only if, W(x) is dense in E, for each $x \in X$.*

COROLLARY 6.4. *Assume that X is a locally compact Hausdorff space. Then $K(X;\mathbb{K}) \otimes E$ is dense in $C_o(X;E)$.*

PROOF. Since X is locally compact $K(X;\mathbb{K})$ is a separating self-adjoint subalgebra of $C_b(X;\mathbb{K})$. The vector subspace $W = K(X;\mathbb{K}) \otimes E$ is a $K(X;\mathbb{K})$ - module such that $W(x) = E$, for all $x \in X$. It remains to apply Corollary 6.3 above.

DEFINITION 6.5. *Let* $W \subset C_o(X;E)$ *be a vector subspace The Stone-Weierstrass hull of* W *in* $C_o(X;E)$ *, denoted by* $\Delta_o(W)$, *is the set* $\Delta(W) \cap C_o(X;E)$.

(For the definition of $\Delta(W)$, see Definition 4.12.

The arguments used in the proof of Lemma 4.16, § 2, Chapter 4, show that

$$\Delta_o(W) = L_A(A \otimes E) = L_A(W)$$

when E is a locally convex Hausdorff space; and $W \subset C_o(X;E)$ is a vector space invariant under composition with elements of $E' \otimes E$.

DEFINITION 6.6. *Let* $W \subset C_o(X;E)$ *be a vector subspace. We say that* W *is a Stone-Weiestrass subspace if* $\Delta_o(W) \subset \overline{W}$, *where the bar denotes the uniform closure of* W *in* $C_o(X;E)$.

THEOREM 6.7. *(Stone Weierstrass) Suppose* E *is a locally convex Hausdorff space. Every self-adjoint polynomial algebra* $W \subset C_o(X;E)$ *is a Stone-Weierstrass subspace, i.e. for every* $f \in C_o(X;E)$, f *belongs to the uniform closure of* W *in* $C_o(X;E)$ *if, and only if:*

(1) for any $x \in X$, *such that* $f(x) \neq 0$ *, there is* $g \in W$ *such that* $g(x) \neq 0$;

(2) for any $x, y \in X$, *with* $f(x) \neq f(y)$, *there is* $g \in W$ *such that* $g(x) \neq g(y)$.

PROOF. By the previous remarks, $\Delta_o(W) = L_A(W) = L_A(A \otimes E)$, where $A = \{\alpha \circ g;\ \alpha \in E',\ g \in W\}$. By Corollary 6.2 applied to the A-module $A \otimes E$, we have $L_A(A \otimes E) = \overline{A \otimes E}$. Since W is polynomial algebra, $\overline{A \otimes E} \subset \overline{W}$. Hence $\Delta_o(W) \subset \overline{W}$.

The converse $\overline{W} \subset \Delta_o(W)$ is true, whenever E is Hausdorff.

COROLLARY 6.8. *Suppose* E *is Hausdorff. Let* $W \subset C_o(X;E)$ *be a self-adjoint polynomial algebra. Then* W *is dense if and only if,* W *is separating and every-where different from zero.*

THEOREM 6.9. *Suppose* E *is a locally convex Hausdorff space.*

Let $W \subset C_0(X;E)$ *be a vector subspace which is invariant under composition with elements of* $E' \otimes E$ *, and let* $A = \{\alpha \circ f;\ \alpha \in E',\ f \in W\}$. *The following conditions are equivalent:*

(1) W *is localizable under* A *in* $C_0(X;E)$.

(2) W *is a Stone-Weierstrass subspace.*

(3) A *is a Stone-Weierstrass subspace.*

PROOF. By previous remark, $\Delta_0(W) = L_A(W)$. Hence (1) and (2) are equivalent.

Assume (2), and let $f \in C_0(X)$ be an element of $\Delta_0(A)$. Let $\varepsilon > 0$ be given. Choose $\alpha \in E'$ with $\alpha \neq 0$, and $v \in E$, with $\alpha(v) = 1$. Let $g = f \otimes v$. Obviously $g \in \Delta_0(W)$. By hypothesis, $g \in \overline{W}$. Let $p \in cs(E)$ be wich that $|\alpha(t)| \leq p(t)$, for all $t \in E$. Let $h \in W$ be chosen so that $p(g(x) - h(x)) < \varepsilon$ for all $x \in X$. Hence $|f(x) - (\alpha \circ h)(x)| < \varepsilon$ for all $x \in X$. Since $\alpha \circ h \in A$, $f \in \bar{A}$, and therefore A is a Stone-Weierstrass subspace.

Finally, assume (3). Since $\bar{A} = \Delta_0(A)$, $\bar{A}$ is a closed self-adjoint subalgebra of $C_0(X)$. Indeed $\Delta_0(A) = \Delta(A) \cap C_0(x)$, and by Proposition 4.15, § 2, Chapter 4, $\Delta(A)$ is a self-adjoint subalgebra of $C(X)$. Let $B = \bar{A}$ .By Corollary 6.2 applied to the B-module $B \otimes E$, we have $L_B(B \otimes E) = \overline{B \otimes E}$. Hence $L_A(W) = L_A(A \otimes E) \subset L_A(B \otimes E) = L_B(B \otimes E) = \overline{B \otimes E} \subset \overline{A \otimes E}$, because $\bar{A} \otimes E \subset \overline{A \otimes E}$. By Lemma 4.1, § 1, Chapter 4, $A \otimes E \subset W$, and therefore $L_A(W) \subset \overline{W}$, which proves (1).

Let Y be a *closed* non-empty subset of a locally compact Hausdorff space X. Then Y is also a locally compact Hausdorff space, and if $f \in C_0(X;E)$ then the restriction $f|Y$ belongs to $C_0(Y;E)$.Let us call T_Y the restriction of $T_Y: C(X;E) \to C(Y;E)$ to the subspace $C_0(X;E)$. Then $T_Y: C_0(X;E) \to C_0(Y;E)$ is a continuous linear map.

LEMMA 6.10. *For any closed non-empty subset* $Y \subset X$, *the linear map* $T_Y: C_0(X;E) \to C_0(Y;E)$ *is a topological homomorphism.*

PROOF. The same proof of Lemma 3.2. applies. Indeed, the set

$F = \{x \in X;\ p(g(x)) \geq \varepsilon\}$ is then *compact* and disjoint from Y.

THEOREM 6.11. *Let* Y *be a closed non-empty subset of a locally compact Hausdorff space* X, *and let* E *be a non-zero Fréchet space. Then* $C_o(X;E)\ |_Y = C_o(Y;E)$.

PROOF. Let $W = C_o(X;E)\ |_Y$. Since X is locally compact, $C_o(X)$ is separating and everywhere different from zero. Hence the same is true of $C_o(X) \otimes E \subset C_o(X;E)$. Taking restrictions to Y and applying Corollary 6.3 (or Corollary 6.4, since $K(X) \subset C_o(X)$), we see that W is dense in $C_o(Y;E)$. We claim that W is closed in $C_o(Y;E)$. Let M be the Kernel of the map T_Y in $C_o(X;E)$. Since T_Y is continuous, M is closed. The space $C_o(X;E)$ is a Fréchet space, because E is Fréchet. The quotient of a Fréchet space by a closed subspace is a Fréchet space. Therefore $C_o(X;E)/M$ is complete. By Lemma 6.10, $C_o(X;E)/M$ and $T_Y(C_o(X;E)) = W$ are linearly topologically isomorphic. Hence W is complete too, and therefore closed in $C_o(X;E)$.

REMARK 6.12. When E is a Banach space, and $f \in C_o(Y;E)$, we can choose $g \in C_o(X;E)$, $g|Y = f$, such that $||f||_Y = ||g||_X$.

REMARK 6.13. Let us now consider a particular case of vector fibrations. Namely, we will consider vector spaces L of cross-sections satisfying the following conditions:

(1) X is a locally compact Hausdorff space;

(2) each E_x is a normed space, whose norm we denote by $t \to ||\ t\ ||$

(3) for each cross-section $f \in L$, the function $x \to$ $\to ||f(x)||$ is upper semi-continuous and vanishes at infinity on X.

In the language of [39], we say in this case that L is a *Nachbin space* of cross-sections, and endow it with the topology of the norm

$$||f|| = \sup \{||f(x)||\ ;\ x \in X\}.$$

The above sup is finite, since the set $\{x \in X;\ ||f(x)|| \geq 1\}$ is compact and the map $x \to ||f(x)||$ is upper semicontinuous. We have then the following strong form of the Stone-Weierstrass Theorem.

THEOREM 6.14. *Let* L *be a Nachbin space of cross-sections satisfying conditions* (1) - (3) *of Remark* 6.13 *and assume that* L *is a* $C_b(X)$ *-module. Then, for every* $C_b(X)$*-submodule* $W \subset L$ *and every* $f \in L$, *we have.*

$$d = \inf_{g\in W} ||f - g|| = \sup_{x\in X} \inf_{g\in W} ||f(x) - g(x)|| = c.$$

PROOF. Clearly, $c \leq d$. To prove the reverse inequality, let $\varepsilon > 0$. For each $x \in X$, there exists $w_x \in W$ such that $||f(x) - w_x(x)|| < c + \varepsilon/2$. Let $U_x = \{t \in X;\ ||f(t) - w_x(t)|| < c+\varepsilon/2\}$ Then U_x is an open subset of X, containing the point x, and its complement in X is compact. By Lemma 5.10, § 2, Chapter 5, applied to the algebra $A = C_b(X;\mathbb{R})$, there exist $x_1,\dots,x_n \in X$, such that to each $\delta > 0$, there correspond functions $a_1,\dots,a_n \in C_b(X;\mathbb{R})$ with $0 \leq a_i \leq 1$, $0 \leq a_i(x) < \delta$ for $x \in X \setminus U_{x_i}$, for $i = 1, 2, \dots, n$, and $a_1 +\dots+ a_n = 1$. Let us choose $\delta > 0$ such that $n M \delta < \varepsilon/2$, where

$$M = \max \{||f - w_i|| ;\ i = 1, 2,\dots n\}$$

and $w_i = w_x$ for $x = x_i$, $i = 1, 2,\dots n$. Consider the corresponding functions $a_1,\dots,a_n$ in $C_b(X;\mathbb{R})$. Let $w = a_1w_1 + \dots + a_n w_n \in W$. We claim that $||f - w|| < c + \varepsilon$. Indeed, given $x \in X$, we have

$$||f(x) - w(x)|| = ||\sum_{i=1}^{n} (f(x) - w_i(x))\, a_i(x)||$$

$$\leq \sum_{i=1}^{n} a_i(x) \cdot || f(x) - w_i(x)||.$$

Now, if $x \notin U_{x_i}$ then $a_i(x) < \delta$, and therefore $a_i(x) \cdot || f(x) - w_i(x)|| < \delta\, || f - w_i|| \leq \delta M$. On the other hand, if $x \in U_{x_i}$, then

$$a_i(x) \cdot || f(x) - w_i(x)|| \leq a_i(x)(c + \varepsilon/2).$$

Hence, $||f(x) - w(x)|| < n\,\delta M + (c + \varepsilon/2) < c + \varepsilon$

This shows that $d < c + \varepsilon$. Since $\varepsilon > 0$ was arbitrary, $d \leq c$, and the proof is seen to be complete.

We shall next consider the question of extreme functionals. Our aim is to prove the theorem of Brosowski, Deutsch and Morris (Lemma 3.3, [10]) generalizing it to vector fibrations satisfying (1), (2) and (3) of Remark 6.13, i.e. to Nachbin spaces of a special kind.

Let then L be a Nachbin space of cross-sections satisfying (1), (2) and (3) of Remark 6.13. Let B' be the unit ball of L', and for each $x \in X$, let B_x' be the unit ball of E_x'. The mapping $\delta_x: E_x' \to L'$ defined by

$$\delta_x(\phi)(f) = \phi(f(x))$$

for all $\phi \in E_x'$ and $f \in L$ is then a continuous linear mapping of norm ≤ 1, and therefore maps B_x' into B'. The map δ_x is one-to-one if $\{f(x); f \in L\} = E_x$. If L is essential (Definition 1.31, § 10, Chapter 1) this happens for all $x \in X$. The image of E_x' by δ_x is contained in the set $\{\psi \in L'; f(x) = 0 \Longrightarrow \psi(f) = 0, \forall f \in L\} = A_x$.

LEMMA 6.15. *Let* L *be a Nachbin space which is an essential* $K(X)$ *-module, satisfying* (1) - (3) *of Remark* 6.13. *The mapping* δ_x *is a linear isometry of* E_x' *onto* A_x *for each* $x \in X$.

PROOF. We saw already that δ_x is a continuous linear mapping of norm ≤ 1, from E_x' into A_x.

Let $\psi \in A_x$. For each $v \in E_x$ choose $f \in L$ such that $f(x) = v$ and put $\varepsilon_x(\psi)(v) = \psi(f)$. Since $\psi \in A_x$, $\varepsilon_x(\psi)$ is well defined on E_x, and it is clearly a linear functional. We claim that $\varepsilon_x(\psi) \in E_x'$. Let $\varepsilon > 0$. Let $v \neq 0$ be given in E_x. Choose $f \in L$ with $f(x) = v$. By condition (3), there is a neighborhood U of x in X, whose complement is compact, and such that, for all $t \in U$, $||f(t)|| < (1 + \varepsilon)\cdot||f(x)||$. Since X is locally compact, there is $g \in K(X)$ such that $0 \leq g \leq 1$, $g(x) = 1$ and $g(t) = 0$ for all $t \notin U$. Since L is a $K(X)$ -module, $g f \in L$ and $||g f|| < (1 + \varepsilon)||f(x)||$. Now $(g f)(x) = v$ and so

$$\begin{aligned}|\varepsilon_x(\psi)(v)| &= |\psi(g\ f)| \leq ||\psi||.||g\ f|| \\ &< ||\psi||\cdot(1+\varepsilon)\cdot||f(x)|| \\ &= ||\psi||.(1+\varepsilon)\cdot||v||\end{aligned}$$

Hence $\varepsilon_x(\psi) \in E'_x$ and $||\varepsilon_x(\psi)|| \leq ||\psi||$ for all $\psi \in A_x$. Therefore $||\varepsilon_x|| \leq 1$. Since δ_x and ε_x are inverses of each other, and of norm ≤ 1, we see that δ_x and ε_x are linear isometries.

LEMMA 6.16. *Let* L *be an essential Nachbin space satisfying conditions* (1) - (3) *of Remark* 6.13. *Let* $Q = \cup \{\delta_x(B'_x); x \in X\}$.

Then (a) Q *is weak*-closed*.

(b) $\overline{co}(Q) = B'$.

PROOF. (a) Suppose $\psi \in L'$ is the weak*-limit of a net $\{\psi_\alpha\}$ in Q. Each ψ_α is of the form $\psi_\alpha = \delta_{x_\alpha}(\phi_\alpha)$, where $x_\alpha \in X$ and $\phi_\alpha \in B'_{x_\alpha}$. Let $Y = X \cup \{w\}$ be the one point compactification of X. We may assume $\{x_\alpha\}$ converges to some point $x \in Y$. If $x \in X$, we proceed as in the proof of Lemma 1.33 (§ 10, Chapter 1) to show that $\psi \in Q$. If $x = w$, we have for any $f \in L$

$$\begin{aligned}|\psi(f)| = \lim |\psi_\alpha(f) &= \lim |\phi_\alpha(f(x_\alpha))| \\ &\leq \lim\sup ||\phi_\alpha||.||f(x_\alpha)|| \\ &\leq \lim\sup ||f(x_\alpha)|| = 0\end{aligned}$$

because $x \to ||f(x)||$ vanishes at w.
Hence $\psi = 0 \in Q$.

(b) The proof of part (b) of Lemma 1.33 (§ 10,Chapter 1) carries over without modification.

THEOREM 6.17. *Let* L *be an essential Nachbin space, which is a* K(X)*-module, satisfying conditions* (1) - (3) *of Remark* 6.13 *then*

$$E(B') = \cup\{\delta_x(E(B'_x)); x \in X, E_x \neq \{0\}\}.$$

COROLLARY 6.18. (*Brosowski, Deutsch, Morris* [10]). *Let* X *be a locally compact Hausdorff space and let* E *be a non-zero normed space. Let* B' *be the unit ball of* $C_0(X;E)$, *and* B'_E *the unit ball of* E'.*Then* $E(B') = \cup\{\delta_x(E(B'_E)); x \in X\}$

PROOF. Apply theorem 6.17 to the Nachbin space $L = C_0(X;E)$.

COROLLARY 6.19. *Let* X *be a locally compact Hausdorff space. Let* B' *be the unit ball of* $C_0(X)'$. *Then*

$$E(B') = \{\lambda\delta_x \; ; \; x \in X, \; \lambda \in \mathbb{K}, |\lambda| = 1\},$$

where $\lambda\delta_x : C_0(X;\mathbb{K}) \to \mathbb{K}$ *is defined by*

$$\lambda\delta_x(f) = \lambda f(x)$$

for all $f \in C_0(X;\mathbb{K})$.

PROOF OF THEOREM 6.17. By Lemma 5, Dunford and Schwartz [20], pg. 440, $E(B')$ Q. Therefore any $\psi \in E(B')$ is of the form $\delta_x(\phi)$ for some $\phi \in B'_x$, $x \in X$. Since δ_x is an isometry, $\phi \in E(B'_x)$.

Conversely, let $\psi \in \delta_x(E(B'_x))$ for some $x \in X$, $E_x \neq \{0\}$. Then $\psi = \delta_x(\phi)$ for some $\phi \in B'_x$. Since $\varepsilon_x(\psi) = \phi$, $\psi \in E(A_x \cap B')$, because ε_x is an isometry.

We claim that $\psi \in E(B')$. Assume, by contradiction, that $\psi \notin E(B')$. Then $\psi = (\psi_1+\psi_2)/2$ for some $\psi_1, \psi_2 \in B'$ and $\psi_1 \neq \psi_2$. Let $f \in L$ with $f(x) = 0$, $||f|| \leq 1$ and $1 > \varepsilon > 0$ be given. Then $U = \{t \in X ; ||f(t)|| < \varepsilon\}$ is open, contains x, and $X \setminus U$ is compact. Choose $g \in K(X)$ with $0 \leq g \leq 1$, $g(x) = 1$ and $g(t) = 0$ for $t \notin U$. Choose $v \in E_x$ such that $||v|| < 1$, $\phi(v)$ is real and $\phi(v) > 1 - \varepsilon$. Choose $h \in L$ with $||h|| \leq 1$ and $h(x) = v$. Let $m = gh$. Since L is $K(X)$-module, $m \in L$. On the other hand $||m|| \leq 1$ and $\phi(m(x)) = \phi(v) > 1 - \varepsilon$. For $t \notin U$, $m(t) = 0$; for $t \in U$, $||f(t)|| < \varepsilon$. Hence $||f + m|| \leq 1 + \varepsilon$. Exactly as in Lemma 1.35, § 10, Chapter 1, one shows that

$$|\psi_1(m) - \psi_2(m)| < 4\sqrt{\varepsilon}$$

and

$$|\psi_1(f + m) - \psi_2(f + m)| < 4\sqrt{\varepsilon},$$

whence

$$|\psi_1(f) - \psi_2(f)| < 8\sqrt{\varepsilon}$$

This proves $\psi_1 - \psi_2 \in A_x$. Therefore $\psi_1 + \psi_2 = 2\psi \in A_x$ implies $\psi_1, \psi_2 \in A_x \cap B'$, which contradicts $\psi \in E(A_x \cap B')$. This ends the proof.

REFERENCES FOR CHAPTER 6.

BROSOWSKI and DEUTSCH [10]

MACHADO and PROLLA [39]

STRÖBELE [63]

CHAPTER 7

THE SPACE $C_b(X;E)$ WITH THE STRICT TOPOLOGY

We start with the Stone-Weierstrass Theorem for algebras and modules. The first such Theorem was obtained by Buck himself (see Buck [11]) : β-density of subalgebras of $C_b(X)$ and β-density of $C_b(X)$-modules in $C_b(X;E)$ when E is finite-dimensional. The latter result was generalized by Wells [67] to the case of any locally convex space E, and subspaces $W \subset C_b(X;E)$ that are A-modules, where $A = \{f \in C_b(X);\ f(x) \subset [0,1]\}$. Further results were obtained by C. Todd (see his Theorem 3, [65]). Our first theorem subsumes all those earlier results.

THEOREM 7.1 *Let X be a locally compact Hausdorff space, and let $A \subset C_b(X;\mathbb{K})$ be a subalgebra. Let $W \subset C_b(X;E)$ be a vector subspace which is an A-module, where E is a locally convex space. Then W is sharply localizable under A in $C_b(X;E)$ equipped with the strict topology β.*

PROOF Apply Corollary 5.29 since $A \subset C_b(X;\mathbb{K})$.

THEOREM 7.2 *Let A and W be as in Theorem 7.1. Assume that A is self-adjoint. Then W is localizable under A in $C_b(X;E)$ with the strict topology β.*

PROOF Since A is self-adjoint, then $G(A) = \mathrm{Re}\,A \cup \mathrm{Im}\,A$ is a strong set of generators for A satisfying the hypothesis of Theorem 5.20, § 3, Chapter 5. On the other hand, since G(A) consists only of real valued functions, $\rho = 2$ and $P_2 = P_A$. Thus W is localizable under A in $C_b(X;E)$ with the strict topology β.

COROLLARY 7.3 *Let A and W be as in Theorem 7.2. Assume that A is separating. Then W is β-dense in $C_b(X;E)$ if, and only if, W(x) is dense in E, for all $x \in X$.*

PROOF For each $x \in X$, there is some $\phi \in C_o(X)$ with $\phi(x) > 0$.

Hence the condition is necessary. The sufficiency follows from Theorem 7.2.

COROLLARY 7.4 *The space* $K(X) \otimes E$ *is* β*-dense in* $C_b(X;E)$.

PROOF Apply Corollary 7.3, with $A = K(X)$ and $W = K(X) \otimes E$.

THEOREM 7.5 *Let* X *and* Y *be two locally compact Hausdorff spaces. Then* $(C_b(X) \otimes C_b(Y)) \otimes E$ *is* β*-dense in the space* $C_b(X \times Y; E)$.

PROOF $A = C_b(X) \otimes C_b(Y)$ is a self-adjoint separating subalgebra of $C_b(X \times Y)$ and $W = A \otimes E$ is such that $W(x) = E$, for each $x \in X$. It remains to apply Corollary 7.3.

In fact the following stronger version of Theorem 7.5 is true.

THEOREM 7.6 *Let* X *and* Y *be two locally compact Hausdorff spaces. Then* $(K(X) \otimes K(Y)) \otimes E$ *is* β*-dense in the space* $C_b(X \times Y; E)$.

PROOF Similar to that of Theorem 7.5.

DEFINITION 7.7 *Let* $W \subset C_b(X;E)$ *be a vector subspace. The Stone-Weierstrass* β*-hull of* W *in* $C_b(X;E)$, *denoted by* $\Delta_\beta(W)$ *is the set* $\Delta(W) \cap C_b(X;E)$.

(For the definition of $\Delta(W)$, see Definition 4.12, §2, Chapter 4).

An obvious modification of the proof of Lemma 4.16, § 2, Chapter 4 shows that

$$\Delta_\beta(W) = L_A(A \otimes E) = L_A(W),$$

when E is a locally convex Hausdorff space, and $W \subset C_b(X;E)$ is a vector subspace invariant under composition with elements of $E' \otimes E$.

DEFINITION 7.8 *Let* $W \subset C_b(X;E)$ *be a vector subspace. We say that* W *is a Stone-Weierstrass subspace if* $\Delta_\beta(W) \subset \overline{W}$, *where the bar denotes the* β*-closure of* W *in* $C_b(X;E)$.

THEOREM 7.9 *(Stone-Weierstrass) Suppose* E *is a locally convex*

Hausdorff space. Every self-adjoint polynomial algebra $W \subset C_b(X;E)$ *is a Stone-Weierstrass subspace. In particular, for every* $f \in C_b(X;E)$, f *belongs to the* β*-closure of* W *in* $C_b(X;E)$ *if, and only if:*

(1) *for any* $x \in X$, *such that* $f(x) \neq 0$, *there is* $g \in W$ *such that* $g(x) \neq 0$;

(2) *for any* $x,y \in X$, *with* $f(x) \neq f(y)$, *there is* $g \in W$ *such that* $g(x) \neq g(y)$.

PROOF By a previous remark, $\Delta_\beta(W) = L_A(W) = L_A(A \otimes E)$, where $A = \{\phi \circ g;\ \phi \in E',\ g \in W\}$. By Theorem 7.2 applied to the A-module $A \otimes E$, we have $L_A(A \otimes E) = \overline{A \otimes E}$. Since W is a polynomial algebra, $\overline{A \otimes E} \subset \overline{W}$. Hence $\Delta_\beta(W) \subset \overline{W}$. The converse, $\overline{W} \subset \Delta_\beta(W)$, is true whenever E is Hausdorff.

COROLLARY 7.10 *Suppose* E *is Hausdorff. Let* $W \subset C_b(X;E)$ *be a self-adjoint polynomial algebra. Then* W *is* β*-dense if, and only if,* W *is separating and everywhere different from zero.*

REMARK For further results and counter-examples see Haydon [30].

THEOREM 7.11 *Suppose* E *is a locally convex Hausdorff space. Let* $W \subset C_b(X;E)$ *be a vector subspace which is invariant under composition with elements of* $E' \otimes E$ *and let* $A=\{\phi \circ f;\ \phi \in E',\ f \in W\}$. *The following conditions are equivalent:*

(1) W *is localizable under* A *in* $C_b(X;E)$.

(2) W *is a Stone-Weierstrass subspace.*

(3) A *is a Stone-Weierstrass subspace.*

PROOF By previous remark, $\Delta_\beta(W) = L_A(W)$. Hence (1) and (2) are equivalent.

Assume (2), i.e. $\Delta_\beta(W) \subset \overline{W}$. Let $f \in C_b(X)$ be an element of $\Delta_\beta(A)$. Let $v \in C_o(X)$, $v \geq 0$, and $\varepsilon > 0$ be given. Choose $\phi \in E'$ and $u \in E$ with $\phi(u) = 1$. Let $g = f \otimes u$. Obviously, $g \in \Delta_\beta(W)$. Let p be a continuous seminorm on E such that $|\phi(t)| \leq p(t)$ for all $t \in E$. By hypothesis, there is $h \in W$ such that $v(x)p(g(x) - h(x)) < \varepsilon$, for all $x \in X$. Hence $v(x)\ |f(x) - (\phi \circ h)(x)| < \varepsilon$, for all $x \in X$. Since $\phi \circ h \in A$,

$f \in \bar{A}$, the β-closure of A in $C_b(X)$, i.e. $\Delta_\beta(A) \subset \bar{A}$.

Finally, assume (3). Since $\bar{A} = \Delta_\beta(A)$, $\bar{A}$ is a β-closed self-adjoint subalgebra of $C_b(X)$. Indeed, $\Delta_\beta(A) = \Delta(A) \cap C_b(X)$, and by Proposition 4.15, § 2, Chapter 4, $\Delta(A)$ is a self-adjoint subalgebra of $C(X)$. Let $B = \bar{A}$. By Theorem 7.2, applied to the polynomial algebra $B \otimes E$, we have $L_B(B \otimes E) = \overline{B \otimes E}$. Hence $L_A(W) = L_A(A \otimes E) \subset L_A(B \otimes E) = L_B(B \otimes E) = \overline{B \otimes E} \subset \overline{A \otimes E}$, because $\bar{A} \otimes E \subset \overline{A \otimes E}$. By Lemma 4.1, § 1, Chapter 4, $A \otimes E \subset W$, and therefore $L_A(W) \subset \bar{W}$, which proves (1).

REMARK From Proposition 4.15, § 2, Chapter 4, and the following facts:

(1) $\Delta_\beta(W) = \Delta(W) \cap C_b(X;E)$; and

(2) β is stronger than the compact-open topology; it follows that $\Delta_\beta(W)$ is the smallest β-closed self-adjoint polynomial algebra contained in $C_b(X;E)$ which contains W.

THEOREM 7.12 *Every proper β-closed $C_b(X)$-module $W \subset C_b(X;E)$ is contained in some proper β-closed $C_b(X)$-module V of codimension one (hence maximal) in $C_b(X;E)$. Moreover, W is the intersection of all maximal proper β-closed $C_b(X)$-modules that contain it.*

PROOF Let $W \subset C_b(X;E)$ be a proper β-closed $C_b(X)$-module. Let $f \in C_b(X;E)$ be a function such that $f \notin W$. Since W is β-closed, and $C_b(X)$ is separating, by Corollary 7.3, there is $x \in X$ such that $f(x) \notin \overline{W(x)}$ in E. By the Hahn-Banach theorem, there is $\phi \in E'$ such that $\phi(f(x)) \neq 0$, while $\phi(g(x)) = 0$ for all $g \in W$. Then $V = \{g \in C_b(X;E);\ \phi(g(x)) = 0\}$ is a β-closed $C_b(X)$-module of codimension one in $C_b(X;E)$, containing W, and $f \notin V$.

COROLLARY 7.13 *All maximal proper β-Closed $C_b(X)$-modules of $C_b(X;E)$ are of the form $\{g \in C_b(X;E);\ \phi(g(x)) = 0\}$ for some $x \in X$ and $\phi \in E'$.*

Before proceeding we need the following elementary properties of $(C_b(X;E), \beta)$ which were proved by Buck [11].

PROPOSITION 7.14 *Let* X *be a locally compact Hausdorff space, let* E *be a locally convex Hausdorff space, and let* β *be the strict topology on* $C_b(X;E)$. *Then*

(1) *if* κ *is the compact-open topology and if* σ *is the uniform topology, then* $\kappa \leq \beta \leq \sigma$;

(2) *the topologies* β *and* σ *have the same bounded sets;*

(3) *on any* σ*-bounded set in* $C_b(X;E)$, *the topology* β *agrees with* κ:

(4) *if* E *is complete, then* $(C_b(X;E),\beta)$ *is complete.*

PROOF (1) Let $\phi \in C_o^+(X)$ and $p \in cs(E)$ be given. For any $f \in C_b(X;E)$ we have

$$||f||_{\phi,p} \leq ||\phi||_X \cdot ||f||_p$$

where $||\phi||_X = \sup \{|\phi(x)| ; x \in X\}$ and $||f||_p = \sup\{p(f(x)); x \in X\}$. This shows that β is smaller than σ. On the other hand, given a compact subset $K \subset X$, choose $\phi \in C_o^+(X)$ such that $\phi(x) = 1$ on K. Then $||f||_{K,p} \leq ||f||_{\phi,p}$ for any $f \in C_b(X;E)$ and $p \in cs(E)$. Therefore, κ is smaller than β.

(2) Since $\beta \leq \sigma$, any uniformly bounded set is strictly bounded. Conversely, if the set $S \subset C_b(X;E)$ is strictly bounded but were not uniformly bounded, choose $f_n \in S$ and $x_n \in X$ such that $p(f_n(x_n)) > n^2$ for a suitable $p \in cs(E)$. Suppose first that $\{x_n\}$ has a convergent subsequence, say $\{x_{n_k}\}$. Let $x_{n_k} \to x$, $x \in X$. Choose $\phi \in C_o^+(X)$ such that $\phi(x_{n_k}) = \phi(x) = 1$ for all $k \in \mathbb{N}$.

Since S is strictly bounded, there exists a constant $M \geq 0$ such that $||f_{n_k}||_{\phi,p} \leq M$ for all $k \in \mathbb{N}$. Hence $p(f_{n_k}(x_{n_k})) \leq M$ for all $k \in \mathbb{N}$, a contradiction to $p(f_{n_k}(x_{n_k})) > n_k$. Therefore, $\{x_n\}$ is discrete and we may choose a sequence of compact sets K_n with $x_n \in K_n$, but the K_n's are pairwise disjoint. Take $\phi_n \in C_o^+(X)$ with range in $[0,1]$, with support contained in K_n, and $\phi_n(x_n) = 1$. Let $\phi(x) = \Sigma\, c_n\, \phi_n(x)$, where $c_n = p(f_n(x_n))^{-1/2}$. Then $\phi \in C_o^+(X)$, and $\phi(x_n) = c_n$, for all $n \in N$. On the other hand $||f_n||_{\phi,p} \geq p(f_n(x_n))^{1/2} \geq n^{1/2}$, contradicting the strict boundedness of S.

(3) Let $S \subset C_b(X;E)$ be a σ-bounded subset. By (1), we have $\kappa|S \leq \beta|S$. Conversely, assume $T \subset S$ is $(\beta|S)$-closed. Let $g \in S$ be in the $(\kappa|S)$-closure of T. Let $p \in cs(E)$, $\varepsilon > 0$ and $\phi \in C_o^+(X)$ be given. Let $M > 0$ be such that $||f||_p \leq M$ for all $f \in S$. Let $K \subset X$ be a compact subset such that $\phi(x) < \varepsilon/2M$ for all $x \notin K$. Choose $f \in T$ such that $||f-g||_{K,p} < \varepsilon/(||\phi||_X + 1)$.

Let $x \in X$. If $x \in K$, then we have

$$\phi(x)p(f(x) - g(x)) < ||\phi||_X \cdot \varepsilon/(||\phi||_X + 1) < \varepsilon.$$

If $x \notin K$, then we have

$$\phi(x)p(f(x) - g(x)) < 2M \cdot \varepsilon/2M = \varepsilon,$$

since both f and g belong to S. Hence $||f-g||_{\phi,p} \leq \varepsilon$, and g belongs to the $(\beta|S)$-closure of T, i.e. $g \in T$ and therefore T is $(\kappa|S)$-closed.

(4) Let $\{f_\alpha\}$ be a net which is β-Cauchy. By (1), $\{f_\alpha\}$ is then κ-Cauchy. Since the space $C(X;E)$ is κ-complete, whenever X is locally compact and E is complete (see Bourbaki, Topologie Générale, Chap. X), $\{f_\alpha\}$ converges in the topology κ to a mapping $f \in C(X;E)$. Let $\phi \in C_o^+(X)$ be given. Let $p \in cs(E)$ and $\varepsilon > 0$ be given. Since $\{f_\alpha\}$ is β-Cauchy, $\{\phi f_\alpha\}$ is σ-Cauchy, and thus converges to a function $g \in C_b(X;E)$ in the topology σ. Since $f_\alpha \to f$ in the topology κ, then $f_\alpha(x) \to f(x)$ for every $x \in X$. Therefore, $g(x) = \phi(x)f(x)$ for all $x \in X$, i.e. $g = \phi\, f$. Notice that each $\phi f_\alpha \in C_o(X;E)$, which is σ-closed in $C_b(X;E)$. Therefore $\phi f \in C_o(X;E)$ for all $\phi \in C_o^+(X)$. The proof of (4) is then complete if we establish the following

LEMMA 7.15 *Let* $f \in C(X;E)$, *and suppose that* $\phi f \in C_o(X;E)$ *for every* $\phi \in C_o^+(X)$. *Then* $f \in C_b(X;E)$.

PROOF (Buck [11]). If $f(X)$ were not bounded in E, then, for some $p \in cs(E)$, there would exist a sequence $\{x_n\}$ in X such that $p(f(x_n)) > n^2$ for all $n \in N$. Since f is continuous, $\{x_n\}$ is discrete and we may choose a sequence of compact sets K_n with $x_n \in K_n$ and the K_n' s are pairwise disjoint. Take $\phi_n \in C_o^+(X)$ with range in $[0,1]$, supp $\phi_n \subset K_n$ and $\phi_n(x_n) = 1$. Let

$$\phi(x) = \Sigma c_n \phi_n(x), \quad \text{where} \quad c_n = p(f(x_n))^{-1}.$$

Then $\phi \in C_o^+(X)$, $\phi(x_n) = c_n$ and therefore $p(\phi(x_n)f(x_n)) = 1$ for all $n \in N$. Thus $\phi f \notin C_o(X;E)$, a contradiction.

REMARK 7.16. The proof that $(C_b(X),\beta)$ has the approximation property was first established by Collins and Dorroh [14]; their argument being a thorough recasting of de Lamadrid's proof for compact X and the uniform topology ([17], pg. 164). When X is completely regular and $C_b(X)$ is equipped with the generalized strict topology T_t, Fremlin, Garling and Haydon([25]), Theorem 10) showed that $(C_b(X),T_t)$ has the approximations property. Their proof is different from and simpler than the proof of Collins and Dorroh. The result of [25] was generalized to $C_b(X;E)$ by Fontenot [24], who considered the case in which E is a normed space with the *metric approximation property*, and $C_b(X;E)$ is equipped with the β_o topology, i.e. the finest locally convex topology on $C_b(X;E)$ which agrees with the compact -open topology on norm bounded sets. It X is locally compact, then $\beta = \beta_o$.

THEOREM 7.17. *Let X be a locally compact Hausdorff space, and let E be a normed space with the metric approximation property, Then $(C_b(X;E),\beta)$ has the approximation property.*

PROOF: See Fontenot [24].

Let E be a locally convex Hausdorff space, and let X and Y be two locally compact Hausdorff spaces. Let $u:Y \to X$ be a continuous mapping. We denote by T_u the linear mapping from $C_b(X;E)$ into $C_b(Y;E)$ defined by composition with u, i.e. $T_u f = f \circ u$ for all $f \in C_b(X;E)$. Let us assume that, for every $\phi \in C_o^+(Y)$, there exists $\psi \in C_o^+(X)$ such that $\phi \leq \psi \circ u$. Then T_u is (β,β)-continuous. Whenever T_u is continuous and $u(Y)$ is closed in X we say that u is *β-admissible*. For example, if $Y \subset X$ is a *closed* subset, and $u : Y \to X$ is the inclusion mapping, it follows from Theorem 6.11 that $C_o^+(X)|Y = C_o^+(Y)$ and therefore u is β-admissible.

THEOREM 7.18 *Let* $u : Y \to X$ *be a β-admissible continuous proper mapping. Then* T_u *is an open mapping for the strict topologies.*

PROOF Let us consider the 0-neighborhood base in $C_b(X;E)$ consisting of all subsets of the form

$$U = \{g \in C_b(X;E);\ \phi(x)p(g(x)) < \varepsilon,\ x \in X\}$$

where $\phi \in C_o^+(X)$, $p \in cs(E)$ and $\varepsilon > 0$. Let

$$W = \{h \in C_b(Y;E);\ \psi(y)p(h(y)) < \varepsilon,\ y \in Y\}$$

where $\psi = \phi \circ u$. We claim that $\psi \in C_o^+(Y)$. Indeed, let $V_\delta = \{y \in Y;\ \phi(u(y)) \geq \delta\}$. If $K_\delta = \{t \in X;\ \phi(t) \geq \delta\}$, then K_δ is compact, and if $y \in V_\delta$ then $u(y) \in K_\delta \cap u(Y)$. Since $u(Y)$ is closed in X, $K = K_\delta \cap u(Y)$ is compact and therefore V_δ is compact, because it is closed and contained in the compact set $u^{-1}(K)$. (Recall that u is a proper mapping). Therefore W is an open β-neighborhood of 0 in $C_b(Y;E)$. Clearly, $T_u(U) \subset W \cap T_u(C_b(X;E))$. Conversely, let $h \in W \cap T_u(C_b(X;E))$. Let $g \in C_b(X;E)$ be such that $h = g \circ u$, i.e. $h(y) = g(u(y))$ for all $y \in Y$. Let $F = \{t \in X;\ \phi(t)p(g(t)) \geq \varepsilon\}$. Then $F \subset X$ is compact and disjoint from $u(Y)$, because $h \in W$. If $F = \emptyset$, then $g \in U$, and therefore $h \in T_u(U)$. If $F \neq \emptyset$, choose $\eta \in K(X)$, $0 \leq \eta \leq 1$, $\eta(x) = 1$ for all $x \in u(Y)$, and $\eta(t) = 0$ for all

$t \in F$. This is possible because $X \backslash u(Y)$ is an open neighborhood of the compact set F, and X is locally compact. Let $f = \eta g \in C_b(X;E)$. Then $h(y) = g(u(y)) = \eta(u(y))g(u(y)) = f(u(y))$ for all $y \in Y$, i.e. $h = T_u(f)$. We claim that $f \in U$. Let $x \in X$. If $x \in F$, then $f(x) = 0$, so $\phi(x)p(f(x)) = 0 < \varepsilon$. If $x \notin F$, then

$\phi(x)p(f(x)) = \phi(x)p(\eta(x)g(x)) = \eta(x)\phi(x)p(g(x)) \leq \phi(x)p(g(x)) < \varepsilon$.

Thus $f \in U$ and $h \in T_u(U)$. Hence $T_u(U) = W \cap T_u(C_b(X;E)$, and $T_u(U)$ is relatively open in $T_u(C_b(X;E))$ for all $U \in$, and T_u is an open mapping, QED.

REMARK For similar results on operators defined on $C_b(X)$ by composition with a continuous mapping between completely regular spaces, when $C_b(X)$ has the generalized strict topology T_t see Theorem 9 and its Corollary, Fremlin, Garling, and Haydon [25].

Let us now consider Bishop's Theorem for the strict topology. When $E = \mathbb{C}$ such a Theorem was proved by Glicksberg [27]. In [51] we proved a version of Bishop's Theorem for Nachbin spaces of vector-valued functions sufficiently general to cover the case of $C_b(X;E)$ with the strict topology β. Here however we will derive it from Theorem 5.20 of Chapter 5.

THEOREM 7.19. *Let* X *be a locally compact Hausdorff space, let* E *be a locally convex Hausdorff space, and let* $A \subset C_b(X;\mathbb{C})$ *be a subalgebra. Let* $W \subset C_b(X;E)$ *be an* A*-module. Then* $f \in C_b(X;E)$ *is in the* β*-closure of* W *if, and only if* $f|K$ *is in the* β*-closure of* $C_b(K;E)$ *for every* A*-antisymmetric set* $K \in \mathcal{K}_A$.

PROOF: Take $G(A) = A$. Then $G(A)$ is a strong set of generators. On the other hand $A \subset C_b(X;E)$, so condition (2) of Theorem 5.20 is also satisfied. Therefore W is sharply localizable under in $C_b(X;E)$. Now $\mathcal{K}_A = P_\rho$ and therefore given $f \in C_b(X;E)$, $\phi \in C_o^+(X)$ and $p \in cs(E)$ then

$$\inf_{g \in W} \|f - g\|_{\phi,p} = \sup_{K \in \mathcal{K}_A} \inf_{g \in W} \|f|K - g|K\|_{\phi,p}$$

In particular, f belongs to the β - closure of W in $C_b(X;E)$ if, and only if, $f|K$ belongs to the β-closure of $W|K$ in $C_b(K:E)$ for any $K \in \mathcal{K}_A$.

Notice that in fact we have proved the "strong" version of Bishop's Theorem.

THEOREM 7.20. *Let* X , E , A *and* W *be as in Theorem 7.22, and Let* $f \in C_b(X;E)$, $\phi \in C_o^+(X)$ *and* $p \in cs(E)$ *be given, then*

$$\inf_{g\in W} \|f - g\|_{\phi,p} = \sup_{K\in\mathcal{K}_A} \inf_{g\in W} \|f|K - g|K\|_{\phi,p}$$

COROLLARY 7.21. *Assume* $(E,\|\cdot\|)$ *is a normed space over* $\mathbb{C}$, *and* X , A *and* W *are as in Theorem 7.23. Then, given* $f \in C_b(X;E)$ *and* $\phi \in C_o^+(X)$, *we have*

$$\inf_{g\in W} \|\phi(f - g)\|_X = \sup_{K\in\mathcal{K}_A} \inf_{g\in W} \|\phi(f - g)\|_K$$

REMARK: In the above statement, if $h \in C_b(S;E)$, where $S \subset X$ is any subset, $\|h\|_S = \sup\{\|h(t)\| ; t \in S\}$. Clearly, the above Corollary implies Bishop's Theorem of Glicksberg [27]. Just take $E = \mathbb{C}$. In fact, the above formula was established by Glicksberg in the case of compact X and the uniform topology in his proof of Bishop's Theorem. See Glicksberg [26], page 419.

REFERENCES FOR CHAPTER 7.

BUCK [11]

COLLINS and DORROH [14]

DE LAMADRID [17]

FONTENOT [24]

FREMLIN, GARLING and HAYDON [25]

GLICKSBERG [26], [27]

HAYDON [30]

PROLLA [51]

TODD [65]

WELLS [67]

CHAPTER 8

THE ε-PRODUCT OF L. SCHWARTZ

§ 1 GENERAL DEFINITIONS

Let E be a locally convex Hausdorff space, with topological dual E'. We denote by E'_c the space E' endowed with the topology of uniform convergence on all absolutely convex compact subsets of E. The space E'_c is a locally convex Hausdorff space, whose topology is defined by the family of seminorms

$$u \in E' \to \sup \{|u(x)|;\ x \in S\}$$

where $S \subset E$ is an absolutely convex compact subset of E. Since the absolutely convex compact subsets of E are, a fortiori, weakly compact, it follows from Mackey's theorem (Grothendieck [28], Corollary 2 to Theorem 7, Chapter II) that the dual $(E'_c)'$ of E'_c is E (as a vector space).

Let now E and F be two locally convex Hausdorff spaces. We shall denote by $\mathcal{L}_e(E'_c;F)$ the vector space of all continuous linear mappings $T : E'_c \to F$ endowed with the topology of uniform convergence on the equicontinuous subsets of E'. The space $\mathcal{L}_e(E'_c;F)$ is then a locally convex Hausdorff space, whose topology is generated by the family of seminorms

$$T \to \sup \{p(T(u));\ u \in V^o\}$$

where p ranges through a system Γ of seminorms defining the topology of F, and V runs through a 0-neighborhood base in E, and we may assume V to be absolutely convex and closed. In fact, we may even assume $V = \{x \in E;\ q(x) \leq 1\}$, where q runs through a system Δ of seminorms defining the topology of E. Indeed, every equicontinuous subset $S \subset E'$ is contained in some V^o with $V = \{x \in E;\ q(x) \leq 1\}$ and $q \in \Delta$.

PROPOSITION 8.1 *The locally convex spaces* $\mathcal{L}_e(E'_c;F)$ *and*

$\mathcal{L}_e(E'_c;F)$ *are linearly topologically isomorphic.*

PROOF Let $T \in \mathcal{L}_e(E'_c;F)$. Its transpose T' is a linear mapping $T' : F' \to (E'_c)'$. We claim that $T' \in \mathcal{L}(F'_c;E)$. Indeed, $(E'_c)' = E$, as a vector space. On the other hand, let $V \subset E$ be an absolutely convex closed neighborhood of 0 in E. To prove continuity of T' we must show that a neighborhood N of 0 in F'_c can be found such that $T'(N) \subset V$. Now the polar $V^o = \{u \in E'; |u(x)| \leq 1, x \in V\}$ is an equicontinuous subset of E', which is weakly compact. Since $\sigma(E',E)$ and the topology of E'_c induce the same topology on the equicontinuous subsets, V^o is a compact subset of E'_c. Therefore $K = T(V^o)$ is an absolutely convex compact subset of F. Its polar K^o is then a neighborhood of 0 in F'_c. Since $T'(K^o) \subset V$, the neighborhood $N = K^o$ is the thing we are looking for.

The transposition mapping $T \to T'$ is therefore a linear isomorphism between $\mathcal{L}(E'_c;F)$ and $\mathcal{L}(F'_c;E)$. We claim that $T \to T'$ is a homeomorphism. By symmetry it is enough to prove continuity.

A 0-neighborhood base in the space $\mathcal{L}_e(F'_c;E)$ is obtained by taking all subsets $U = \{f; f(S) \subset W\}$, when S runs through all equicontinuous subsets of F' and W runs through a 0-neighborhood base in E. We may assume W to be absolutely convex and closed. If $S \subset F'$ is equicontinuous, there exists an absolutely convex closed neighborhood V of 0 in F such that $S \subset V^o$. On the other hand, W^o is an equicontinuous subset of E'. Therefore $N = \{T; T(W^o) \subset V\}$ is a neighborhood of 0 in $\mathcal{L}_e(E'_c;F)$. Since $T(W^o) \subset V$ implies $T'(V^o) \subset W^{oo}$, and $W^{oo} = W$, we see that $T \in N$ implies $T' \in U$, i.e. $T \to T'$ is continuous.

DEFINITION 8.2 ([59]) *We define the ε-product of* E *and* F *by setting*

$$E \varepsilon F = \mathcal{L}_e(F'_c;E).$$

By the above Proposition 1, we may identify $E \varepsilon F$ with $\mathcal{L}_e(E'_c;F)$, i.e., $E \varepsilon F$ and $F \varepsilon E$ are linearly topologically isomorphic.

REMARK. When E is quasi-complete (i.e., when the closed bounded sets in E are complete), then E'_c has the topology of uniform convergence on *all* compact sets of E. Indeed, in a quasi-complete space, the closed absolutely convex hull of a compact set is compact.

PROPOSITION 8.3 *If* E *and* F *are quasi-complete (resp. complete), then the* ε*-product* E ε F *is quasi-complete (resp. complete).*

PROOF See Schwartz [59]. We now show that we may identify $E \otimes_\varepsilon F$ with a subspace of E ε F. To this end we first recall the definition of $E \otimes_\varepsilon F$.

If E and F are vector spaces over 𝕂, then B(E;F) denotes the vector space of all bilinear forms on E × F. The mapping $f \to f(x,y)$, for each pair (x,y) in E × F is then a linear form on B(E,F), i.e. an element of the algebraic dual $B(E,F)^*$ of B(E;F). This linear form is denoted by $x \otimes y$. The mapping defined by $\chi(x,y) = x \otimes y$ is then a bilinear mapping from E × F into $B(E,F)^*$. The linear span of $\chi(E \times F)$ in $B(E,F)^*$ is called the *tensor product* of E and F, and is denoted by $E \otimes F$. Each element $u \in E \otimes F$ is a finite sum of the form

$$u = \sum_{i=1}^{r} x_i \otimes y_i$$

with $x_i \in E$, $y_i \in F$, $i = 1,2,\ldots,r$. This representation is not unique, but we can assume that $\{x_i\}$ and $\{y_i\}$ are linearly independent in E and F respectively. The number r is then uniquely determined and it is called the *rank* of the element $u \in E \otimes F$.

There are several useful topologies on $E \otimes F$, when E and F are locally convex Hausdorff spaces. We are interested here in the topology τ_e of *bi-equicontinuous convergence.* We identify each element of $E \otimes F$ with a *linear form* on $E' \otimes F'$ by means of the formula

$$(1) \qquad (x \otimes y)(x' \otimes y') = x'(x)\, y'(y)$$

extended by linearity. The topology τ_e is the topology of uniform convergence on the sets of the form $\chi(S \times T)$, where S and T run through the equicontinuous subsets of E' and F' respectively.

Another way of characterizing τ_e is the following.

Each element $x \otimes y$, by means of formula (1) defines a *bilinear form* on $E'_\sigma \times F'_\sigma$ which is separately continuous, i.e., we can identify $E \otimes F$ with a subspace of $\mathcal{B}(E'_\sigma, F'_\sigma)$, the vector space of all bilinear forms on $E'_\sigma \times F'_\sigma$ which are separately continuous. The topology τ_e is then the topology induced on $E \otimes F$ by the $\mathfrak{S} \times \mathfrak{T}$ -topology, where $\mathfrak{S}$ and $\mathfrak{T}$ are the families of equicontinuous subsets of E' and F' respectively.

If $u \in E \otimes F$, say $u = \sum_{i=1}^{r} x_i \otimes y_i$, let us define $\tilde{u} : F' \to E$ by

$$\tilde{u}(y') = \sum_{i=1}^{r} y'(y_i)x_i$$

for all $y' \in F'$. The mapping $\tilde{u}$ is obviously linear and does not depend on the particular representation of u. We claim that the map $\tilde{u}$ belongs to $\mathcal{L}(F'_c;E)$. Indeed, if the net $y'_\alpha \to 0$ in F'_c, then $y'_\alpha(y_i) \to 0$ for all $i = 1,2,\ldots,r$. Hence $\tilde{u}(y') \to 0$ in E. The mapping $u \to \tilde{u}$ is then a linear one-to-one mapping from $E \otimes F$ onto a subspace of $\mathcal{L}(F'_c;E)$. We shall denote by $E \otimes_\varepsilon F$ the image of $E \otimes F$ in $\mathcal{L}_e(F'_c;E) = E \,\varepsilon\, F$, with the induced topology. Since the topology of E is the topology of uniform convergence on the equicontinuous sets of E', the topology induced by $E \,\varepsilon\, F$ on $E \otimes F$ is the topology τ_e of bi-equicontinuous convergence. The completion of $E \otimes_\varepsilon F$ will be denoted by $E \hat{\otimes}_\varepsilon F$, and it is called the *injective* tensor product.

§ 2 SPACES OF CONTINUOUS FUNCTIONS

In this section we establish a representation theorem for the ε-product of C(X) and E, when X is a $k_{\mathbb{R}}$-space and E is a quasi-complete (resp. complete) Hausdorff space. Before proceeding, we recall the definition of a $k_{\mathbb{R}}$-space.

DEFINITION 8.4 *A Hausdorff space X is said to be a $k_{\mathbb{R}}$ -space if, for every function $f : X \to \mathbb{R}$ such that $f|K$ is continuous, for each compact subset $K \subset X$, the function f itself is continuous.*

We mention that, when X is a $k_{\mathbb{R}}$-space, and Y is a completely regular Hausdorff space, and $f : X \to Y$ is such that $f|K$ is continuous, for each compact subset $K \subset X$, then $f \in C(X;Y)$.

The following result shows the equivalence between the completeness of $C(X;\mathbb{K})$ endowed with the compact-open topology and the property of X being a $k_{\mathbb{R}}$-space.

THEOREM 8.5 *Let X be a completely regular Hausdorff space. The following conditions are equivalent.*

(a) $C(X;\mathbb{K})$ *is complete under the compact-open topology.*

(b) $C(X;\mathbb{K})$ *is quasi-complete under the compact-open topology.*

(c) X *is a* $k_{\mathbb{R}}$*-space.*

PROOF See Warner [66], Theorem 1.

THEOREM 8.6 *Let X be a completely regular Hausdorff space, which is a* $k_{\mathbb{R}}$*-space, and let E be a quasi-complete locally convex Hausdorff space. Then* $C(X) \varepsilon E$ *and* $C(X;E)$ *are linearly topologically isomorphic. If, moreover E is complete,* $C(X) \hat{\otimes}_{\varepsilon} E$ *and* $C(X) \varepsilon E$ *are linearly topologically isomorphic.*

We first prove the following lemma.

LEMMA 8.7 *If X is a (completely regular) Hausdorff space, which is a* $k_{\mathbb{R}}$*-space, the mapping* $\Delta : x \to \delta_x$ *is a continuous mapping from X into* $C(X;\mathbb{K})'_c$.

PROOF Since each one-point set $\{x\}$ is compact, the map $\delta_x : f \to f(x)$ belongs to $C(X;\mathbb{K})'$. Let us write $F = C(X;\mathbb{K})$. By the definition of the weak*-topology $\sigma(F',F)$, the map $\Delta: X \to F'_\sigma$ is always continuous. Let $K \subset X$ be a compact subset, and let p_K be the seminorm $f \in F \to \sup\{|f(x)|;\ x \in K\}$. Then $\Delta(K) \subset \{f \in F;\ p_K(f) \leq 1\}^{\circ}$. Therefore, Δ maps compact sets into equicontinuous sets. On these F'_σ and F'_c induce the same topology. Hence $\Delta|K$ is continuous as a map from K into F'_c, for each compact set $K \subset X$. By the remark made after Definition 1, Δ is a continuous mapping from X into F'_c.

PROOF OF THEOREM 8.6 As in the proof of Lemma 8.7, let us write $F = C(X;\mathbb{K})$. Define $\phi : \mathcal{L}_e(F'_c;E) \to G$, by $\phi(T) = T \circ \Delta$, where we have defined $G = C(X;E)$. We claim that ϕ is injective. When $E = \mathbb{K}$, this follows from

$$\mathcal{L}(F'_c;\mathbb{K}) = (F'_c)' = F = C(X;\mathbb{K}).$$

For the case of a general E, notice that $\phi(T) = 0$ implies $u \circ (T \circ \Delta) = 0$ for all $u \in E'$. Hence $(u \circ T) \circ \Delta = 0$, for all $u \in E'$. By the previous case, $u \circ T = 0$, for all $u \in E'$. Therefore $T = 0$. We now make the following

CLAIM The map ϕ is onto $C(X;E) = G$.

PROOF OF CLAIM Let $g \in C(X;E)$. For each $u \in E'$, consider $u \circ g \in C(X;\mathbb{K})$. The linear map $T : E'_c \to C(X;\mathbb{K})$ so defined is continuous. Indeed, given $K \subset X$ compact, let

$V = \{f \in C(X;\mathbb{K});\ |f(x)| \leq 1$, for all $x \in K\}$. Since g is continuous, $g(K)$ is compact. Let K_1 be the absolutely convex closed hull of $g(K)$ in E. Since E is quasi-complete, K_1 is compact. On the other hand, $T(K_1^0) \subset V$. Therefore T is continuous. Since it is obviously linear, $T \in \mathcal{L}(E'_c;C(X;\mathbb{K}))$. Then, its transpose T' belongs to $\mathcal{L}(F'_c;E)$. To prove that $\phi(T') = g$, notice that for every $x \in X$ and $u \in E'$, we have

$$< (T' \circ \Delta)(x), u > = < \Delta(x), T(u)> =$$

$$= < \delta_x, T(u) > = T(u)(x) = u(g(x)).$$

It then follows that $\phi(T')(x) = g(x)$ for all $x \in X$, i.e., $\phi(T') = g$. Thus ϕ is onto G, and this completes the proof of the claim.

To finish the proof of the Theorem, we must show that ϕ is a homeomorphism. Indeed for any net (T_α) in the space $\mathcal{L}_e(F'_c;E)$ the following are equivalent statements:

(1) The net $T_\alpha \to 0$ in $\mathcal{L}_e(F'_c;E)$.

(2) The net $T'_\alpha \to 0$ in $\mathcal{L}_e(E'_c;F)$.

(3) $T'_\alpha u \to 0$ in F, uniformly in $u \in S$, for each equicontinuous subset $S \subset E'$.

(4) $(T_\alpha \circ \Delta)(x) \to 0$ in E, uniformly in $x \in K$, for each compact subset $K \subset X$.

(5) $\phi(T_\alpha) \to 0$ in $G = C(X;E)$.

Since $\mathcal{L}_e(F'_c;E)$ is by Proposition 8.1, §1, linearly topologically isomorphic to the space $F \varepsilon E = C(X) \varepsilon E$, this completes the proof of the first part of Theorem 2.

Assume now that E is complete. Since X is a $k_{\mathbb{R}}$-space C(X) is complete too, by Theorem 8.5. It then follows from Proposition 8.3 that $C(X) \varepsilon E$ is complete. Now when we identify $C(X) \varepsilon E$ and $C(X;E)$, the vector subspace $C(X) \otimes_\varepsilon E \subset C(X) \varepsilon E$ is identified with the set of functions $f \in C(X;E)$ such that $f(X)$ is contained in some finite-dimensional subspace of E, i.e., with the space of all finite sums of functions of the form $x \to g(x)v$, where $g \in C(X)$ and $v \in E$. By Theorem 1.14, §6, Chapter 1, this space is dense in $C(X;E)$. We have seen that $C(X;E)$ is complete, therefore

$$C(X) \hat{\otimes}_\varepsilon E = C(X) \varepsilon E.$$

This completes the proof.

§ 3 THE APPROXIMATION PROPERTY

We recall that a locally convex space E has the *approximation property* if the identity map e can be approximated, uniformly on every compact set in E, by continuous linear maps of finite rank. In [22], Enflo has shown that there is a Banach space which fails the approximation property. For an account of the approximation property on function spaces, in particular in Nachbin spaces, see the papers of Bierstedt [6] and Bierstedt and Meise [7].

The following result is due to L. Schwartz. The proof of (3) ⇒ (2) given below follows Schaefer [55], Chapter III, §9, Proposition 9.2.

THEOREM 8.7 *Let* E *be a quasi-complete locally convex Hausdorff space. Then the following are equivalent.*

(1) E *has the approximation property.*

(2) $E \otimes_\varepsilon F$ *is dense in* $E \varepsilon F$, *for all locally convex*

spaces F.

(3) $E \otimes_\varepsilon F$ *is dense in* $E\,\varepsilon\,F$, *for all Banach spaces* F.

PROOF (1) ⟹ (2) (Schwartz [59]).

Let $T \in \mathcal{L}_e(F'_c;E) = E\,\varepsilon\,F$. Let $\mathcal{L}_c(E)$ denote the space of all continuous linear maps from E into E with the topology of compact convergence. The mapping $\Theta : v \to v \circ T$ from $\mathcal{L}_c(E)$ into $\mathcal{L}_e(F'_c;E)$ is continuous, since T(S) is a relatively compact subset of E, for every equicontinuous subset $S \subset F'$. To see this, notice that the weak*-closure $\overline{S}$ of S is equicontinuous too, $\overline{S}$ is weak*-compact, and on $\overline{S}$ the topologies of F'_c and F'_σ coincide. Hence $\overline{S}$ is compact in F'_c, and S is relatively compact in F'_c. Since $T \in \mathcal{L}(F'_c;E)$, T(S) is relatively compact in E. Now, if $v \in E' \otimes E$, then $v \circ T \in F \otimes_\varepsilon E$, because $(F'_c)' = F$. On the other hand, the identity map e on E is such that $\Theta(e) = T$. Hence, if e is in the closure of $E' \otimes E$ in the space $\mathcal{L}_c(E)$, then T is in the closure of $F \otimes_\varepsilon E$ in $\mathcal{L}_e(F'_c;E) = E\,\varepsilon\,F$.

(2) ⟹ (3). Obvious.

(3) ⟹ (2). Let F be a locally convex space. Let B be a 0-neighborhood base of absolutely convex closed sets in F. For each $V \in B$, let $\phi_V : F \to \tilde{F}_V$ denote the canonical map (Schaefer [57], pg. 97). Let $S \subset E'$ be an equicontinuous set, and let $V \in B$ be given. Let us write $\phi = \phi_V$ and $G = \tilde{F}_V$. Let $W = (1/4)\,\overline{\phi(V)} \subset G$. Then $\phi^{-1}(W) \subset V$. Since $\phi(F)$ is dense in G, and by hypothesis $E \otimes_\varepsilon G$ is dense in the space $E\,\varepsilon\,G$, it follows that $E \otimes_\varepsilon \phi(F)$ is dense in $E\,\varepsilon\,G$. Hence, given $T \in E\,\varepsilon\,F = \mathcal{L}_e(E'_c;F)$, then $\phi \circ T \in \mathcal{L}_e(E'_c;G)$. Therefore, we can find $w \in E \otimes_\varepsilon \phi(F)$ such that $w(x) - (\phi \circ T)(x) \in W$ for all $x \in S$. Suppose $w = \sum_{i=1}^{r} x_i \otimes \phi(y_i)$. Then $\phi\left(\sum_{i=1}^{r} x(x_i)y_i - T(x)\right) \in W$, for all $x \in S$, and then $\sum_{i=1}^{r} x(x_i)y_i - T(x) \in V$, for all $x \in S$. Let $v = \sum_{i=1}^{r} x_i \otimes y_i$. Then $v \in E \otimes_\varepsilon F$, and $v(x) - T(x) \in V$, for all

$x \in S$. Since S and V were arbitrary, $E \otimes_\varepsilon F$ is dense in the space $\mathcal{L}_e(E'_c;F) = E \varepsilon F$.

(2) $\Longrightarrow$ (1). (Schwartz [59]).

Take $F = E'_c$. By the Corollary to Proposition 5, Schwartz [59], $\mathcal{L}_c(E)$ is isomorphic to a subspace of $E \varepsilon E'_c$. Since $E \otimes_\varepsilon E' \subset \mathcal{L}_c(E) \subset E \varepsilon E'_c$, and by hypothesis, $E \otimes_\varepsilon E'$ is dense in $E \varepsilon E'_c$, it follows that $E \otimes_\varepsilon E'$ is dense in $\mathcal{L}_c(E)$, and therefore E has the approximation property.

COROLLARY 8.8 *Let* X *be a completely regular Hausdorff space, which is a* $k_{\mathbb{R}}$*-space. Then* $C(X;\mathbb{K})$ *equipped with the compact-open topology has the approximation property.*

PROOF For every Banach space E, by Theorem 8.6, § 2, the following are isomorphic spaces: $C(X) \hat{\otimes}_\varepsilon E$, $C(X) \varepsilon E$, and $C(X;E)$.

COROLLARY 8.9 *For every compact Hausdorff space* X, *the Banach space* C(X) *has the approximation property.*

§ 4 MERGELYAN'S THEOREM

In this section we shall prove a vector-valued version of Mergelyan's Theorem. Let $K \subset \mathbb{C}$ be a compact subset such that $\mathbb{C}\setminus K$ is connected. For every complete locally Hausdorff space E over $\mathbb{C}$, let $A(K;E)$ denote the closed subspace of $C(K;E)$ of all those $f \in C(X;E)$ which are holomorphic on the interior of K. Mergelyan's Theorem states that $A(K;\mathbb{C})$ is the closure in $C(K;\mathbb{C})$ of all polynomials with complex coefficients. (Rudin [55], Theorem 20.5). We shall prove a vector-valued version of this result, due independently to Bierstedt [5], and Briem, Laursen, and Pedersen [9]. We shall present Bierstedt's proof.

We begin with the following result (proved by Bierstedt for Nachbin spaces) which is the key to the relation between the approximation property for subspaces of C(X) and subspaces of C(X;E).

THEOREM 8.10 *Let* X *be a completely regular Hausdorff space, which is a* $k_{\mathbb{R}}$*-space, let* $Y \subset C(X)$ *be a closed subspace, and let* E *be a complete locally convex Hausdorff space. Then* $Y \varepsilon E$ *is linearly topologically isomorphic with the vector subspace of all* $f \in C(X;E)$ *such that* $u \circ f \in Y$, *for all* $u \in E'$.

PROOF We first remark that, when E, F and G are three locally convex Hausdorff spaces, and F is a topological vector subspace of G, then $F \varepsilon E$ is identified with a subspace of $G \varepsilon E$, i.e., $F \varepsilon E \subset G \varepsilon E$ topologically. From this it follows that, $Y \varepsilon E$ is isomorphic with a subspace of $C(X) \varepsilon E = C(X;E)$. Let $W = \{f \in C(X;E);\ u \circ f \in Y, \text{ for all } u \in E'\}$. Then $u \circ f \in (Y_c')' = Y$, for all $f \in Y \varepsilon E$, and $u \in E'$. Hence $Y \varepsilon E \subset W$. Conversely, if $f \in W$, the mapping $u \to u \circ f$ maps E_c' into Y, i.e. $f \in Y \varepsilon E$.

COROLLARY 8.11 *Let* X *and* Y *be as in Theorem* 8.10. *The following are equivalent.*

(1) Y *has the approximation property.*

(2) *For all complete locally convex Hausdorff spaces* E, $Y \otimes E$ *is dense in* $\{f \in C(X;E); u \circ f \in Y$, *for all* $u \in E'\}$.

(3) *For all Banach spaces* E, $Y \otimes E$ *is dense in* $\{f \in C(X;E);\ u \circ f \in Y$, *for all* $u \in E'\}$.

We can now prove the vector-valued version of Mergelyan's Theorem.

THOEREM 8.12 *If* $K \subset \mathbb{C}$ *is a compact subset which has a connected complement, and* E *is a complete locally convex Hausdorff space over* $\mathbb{C}$, *then* $A(K;E)$ *is the closure in* $C(K;E)$ *of* $\mathcal{P}(\mathbb{C}) \otimes E$.

PROOF Let $Y = A(K;\mathbb{C})$. Since holomorphy and weak holomorphy coincide, $A(K;E) = \{f \in C(K;E);\ u \circ f \in Y, \text{ for all } u \in E'\}$. The space $A(K;\mathbb{C})$ has the approximation property (see [21]). Hence, by the previous Corollary 8.11, $A(K;\mathbb{C}) \otimes E$ is dense in $A(K;E)$. By Mergelyan's Theorem, the set $\mathcal{P}(\mathbb{C})|K$ is dense in $A(K;\mathbb{C})$. Therefore $(\mathcal{P}(\mathbb{C}) \otimes E)|K$ is dense in $A(K;\mathbb{C}) \otimes E$, and hence it is dense in $A(K;E)$. Notice that the functions of $\mathcal{P}(\mathbb{C}) \otimes E$ are of the form $Z \to \sum_{i=0}^{n} Z^i x_i$, $n \in \mathbb{N}$, $x_i \in E$, $i = 0,1,2,\ldots,n$.

Let us consider now the case of holomorphic functions on open subsets $U \subset \mathbb{C}^n$. If E is a complete locally convex Hausdorff space over $\mathbb{C}$, then $\mathcal{H}(U;E)$ denotes the set of all holomorphic E-valued functions on U, endowed with the compact-open topology. When $E = \mathbb{C}$, we write simply $\mathcal{H}(U)$.

THEOREM 8.13 *Let U be an open non-void subset of $\mathbb{C}^n$, and let E be a complete locally convex Hausdorff space over $\mathbb{C}$. Then $\mathcal{H}(U) \hat{\otimes}_\varepsilon E$, $\mathcal{H}(U) \varepsilon E$ and $\mathcal{H}(U;E)$ are linearly topologically isomorphic.*

PROOF We first remark that $\mathcal{H}(U;E) = \{f \in C(U;E)\ u \circ f \in \mathcal{H}(U)$, for all $u \in E'\}$. By Theorem 8.10, $\mathcal{H}(U) \varepsilon E$ is linearly topologically isomorphic with $\mathcal{H}(U;E)$. On the other hand, when we identify $C(U) \varepsilon E$ and $C(X;E)$, the vector subspace $\mathcal{H}(U) \otimes_\varepsilon E \subset \mathcal{H}(U) \varepsilon E$ is identified with the set of functions $f \in \mathcal{H}(U;E)$ such that $f(U)$ is contained in some finite-dimensional subspace of E, i.e., with the space of all finite sums of functions of the form $x \to g(x)v$, where $g \in \mathcal{H}(U)$ and $v \in E$. This spaces is dense in $\mathcal{H}(U;E)$ (see Grothendieck [25]). Since the latter space is complete, we have

$$\mathcal{H}(U) \hat{\otimes}_\varepsilon E = \mathcal{H}(U;E).$$

This completes the proof.

COROLLARY 8.14 *Let U be an open non-void subset of $\mathbb{C}^n$. Then $\mathcal{H}(U)$ has the approximation property, when equipped with the compact-open topology.*

In Chapter 4, § 1, we defined the space $\mathcal{H}(E)$ of all holomorphic functions $f : E \to \mathbb{C}$ defined on a complex Banach space E. The following result of Aron and Schottenloher [3] shows the equivalence between the approximation property for E and for $\mathcal{H}(E)$ with the compact-open topology.

THEOREM 8.15 *Let E be a complex Banach space. The following are equivalent:*

(1) *E has the approximation property.*

(2) *$\mathcal{H}(E)$ endowed with the compact-open topology has the approximation property.*

PROOF (1) $\Longrightarrow$ (2). Since E is a $k_{\mathbb{R}}$-space, and $\mathcal{H}(E)$ is complete, hence closed in $C(E)$, by Corollary 8.11, all that we have to prove is that $\mathcal{H}(E) \otimes F$ is dense in the set $W = \{f \in C(E;F);\ u \circ f \in \mathcal{H}(E)$, for all $u \in F'\}$ for all Banach spaces F. However, since E is a Banach space, $W = \mathcal{H}(E;F)$. Let then $f \in \mathcal{H}(E;F)$, $K \subset E$ compact and $\varepsilon > 0$ be given. By uniform continuity of f on K, there exists a $\delta > 0$ such that $x \in K$, $y \in E$, $||x-y|| < \delta$ imply $||f(x) - f(y)|| < \varepsilon/2$. By the approximation property, there exists $u \in E' \otimes E$ such that $x \in K$ implies $||x-u(x)|| < \delta$. We next remark that $u(E)$ is finite-dimensional and $f|u(E)$ belongs to the space $\mathcal{H}(u(E);F)$. Since $\mathcal{H}(u(E))|\otimes F$ is dense in $\mathcal{H}(u(E);F)$, there exists $g \in \mathcal{H}(u(E);F)$ such that $||g(t) - f(t)|| < \varepsilon/2$ for all $t \in u(K)$. Let $h = g \circ u$. Then $h \in \mathcal{H}(E) \otimes F$, and for all $x \in K$ we have $||f(x) - h(x)|| < \varepsilon$.

(2) $\Longrightarrow$ (1). Since E has the approximation property if, and only if, E'_c has the approximation property, and since E'_c is a complemented subspace of $\mathcal{H}(E)$, then (2) $\Longrightarrow$ (1) follows from the fact that a complemented subspace of a space with the approximation property has the approximation property.

§ 5 LOCALIZATION OF THE APPROXIMATION PROPERTY

The results of this section are due to Bierstedt [6], who derived a "localization" of the approximation property for closed subspaces of certain Nachbin spaces. We will consider only the case of $C(X)$ for X compact.

THEOREM 8.16 *Let X be a compact Hausdorff space, let $A \subset C(X)$ be a subalgebra, and let $W \subset C(X)$ be a closed A-module. If $W|K \subset C(K)$ has the approximation property, for each maximal A-antisymmetric set $K \subset X$, then W has the approximation property.*

PROOF By Corollary 8.11, § 4, we have to prove that, for each complete locally convex Hausdorff space E, the A-module $W \otimes E$ is dense in $\{f \in C(X;E);\ u \circ f \in W \text{ for all } u \in E'\} = S$. Let $K \subset X$ be a maximal antisymmetric set for A, and let $T = \{g \in C(K;E);\ u \circ g \in W|K, \text{ for all } u \in E'\}$. Since $W|K$ is closed and has the approximation property, it follows from Corollary 8.11 that $(W|K) \otimes E$ is dense in T. However, $S|K \subset T$ and $(W \otimes E)|K = (W|K) \otimes E$. Hence $(W \otimes E)|K$ is dense in $S|K$, for each maximal antisymmetric set $K \subset X$. By Theorem 1.27, § 8, Chapter 1, $W \otimes E$ is dense in S.

COROLLARY 8.17 *Let X be a compact Hausdorff space. Every closed ideal* $I \subset C(X)$ *has the approximation property.*

For the next example, let E be a locally compact Hausdorff space, and let $U \subset \mathbb{C}^n$, $n \geq 1$, be an open non-void subset. For $\Omega \subset U \times E$, open and non-void too, define for each $x \in E$, the "slice" $\Omega_x = \{Z \in U;\ (Z,x) \in \Omega\}$. Then, Ω_x is an open subset of U. We define $C\mathcal{H}(\Omega) = \{f \in C(\Omega);\ Z \to f(Z,x) \text{ belongs to } \mathcal{H}(\Omega_x)$ for each $x \in E$ such that $\Omega_x \neq \phi\}$, equipped with the compact-open topology. Then $C\mathcal{H}(\Omega)$ is a closed subspace of $C(\Omega)$; in fact, it is a closed A-module, where A is the algebra $\{f \in C(\Omega);\ f$ is constant on $\Omega_x \times \{x\}$, for each $x \in E\}$. The maximal antisymmetric sets for A are the sets of the form $\Omega_x \times \{x\}$, for each $x \in E$ such that $\Omega_x \neq \phi$. If $Y = C\mathcal{H}(\Omega)$ and $K = \Omega_x \times \{x\}$, we may identify $Y|K$ with a subspace of $\mathcal{H}(\Omega_x)$. Since $\mathcal{H}(\Omega_x)$ is nuclear, $Y|K$ is nuclear too. Hence $Y|K$ has the approximation property. By Theorem 1 above Y has the approximation property. We have thus proved the following

THEOREM 8.18 $C\mathcal{H}(\Omega)$ *has the approximation property.*

Let $K \subset \mathbb{C} \times E$ be a non-empty closed subset such that $K_x = \{Z \in \mathbb{C};\ (Z,x) \in K\}$ is a compact subset of $\mathbb{C}$. Define $CA(K) = \{f \in C(K);\ Z \to f(Z,x)$ is analytic on the interior of K_x, for each $x \in E$ such that the interior of K_x is $\neq \phi\}$. We further assume that, for each $x \in E$, the complement of K_x in $\mathbb{C}$ is connected.

THEOREM 8.19 CA(K), *under the above hypothesis, has the approximation property.*

PROOF Let Y = CA(K). As a subspace of C(K) with the compact-open topology, Y is closed. Moreover, Y is an A-module, where A is the algebra $\{f \in C(K);\ f$ is constant on $K_x \times \{x\},\ x \in E\}$. As before, the maximal antisymmetric sets for A are the sets $K_x \times \{x\}$, with $K_x \neq \phi$. For each such x we may identify $Y|X$, where $X = K_x \times \{x\}$, with a subspace of $A(K_x) = \{f \in C(K_x);\ f$ is analytic on the interior of $K_x\}$. By Mergelyan's Theorem, since $\mathbb{C}\backslash K_x$ is connected, the polynomials are dense in $A(K_x)$. Since Y contains the polynomials, i.e. the functions of the form $(Z,x) \to p(Z)$, where p is a polynomial, $Y|K$ is dense in $A(K_x)$. Since $A(K_x)$ has the approximation property (Eifler [21]), $Y|K$ has the approximation property.

REFERENCES FOR CHAPTER 8.

ARON and SCHOTTENLOHER [3]

BIERSTEDT [5], [6]

BIERSTEDT and MEISE [7]

BRIEM, LAURSEN and PEDERSEN [9]

EIFLER [21]

ENFLO [22]

RUDIN [55]

SCHAEFER [57]

SCHWARTZ [59]

WARNER [66]

CHAPTER 9

NONARCHIMEDEAN APPROXIMATION THEORY

§ 1. VALUED FIELDS

DEFINITION 9.1. *Let* F *be a field. A (rank one or real-valued) valuation of* F *is a mapping* $|\cdot| : F \to \mathbb{R}$ *satisfying the following conditions:*

(1) $|x| \geq 0$, *for all* $x \in F$;

(2) $|x| = 0$, *if and only if*, $x = 0$;

(3) $|xy| = |x| \cdot |y|$, *for all* $x, y \in F$;

(4) $|x + y| \leq |x| + |y|$, *for all* $x, y \in F$.

If $|\cdot|$ is a valuation of F, we say that $(F, |\cdot|)$ is a *valued field* or a *field with valuation.*

Any field F can be provided with a valuation, namely the *trivial* valuation, defined as follows: $|x| = 1$ for all

$$x \in F, x \neq 0, \quad \text{and} \quad |x| = 0 \text{ if } x = 0.$$

The field $\mathbb{K}$ ($\mathbb{K} = \mathbb{R}$ or $\mathbb{C}$) with its usual absolute value is another example of a valued field.

DEFINITION 9.2. *Let* $(F, |\cdot|)$ *be a valued field. We say that* $|\cdot|$ *is nonarchimedean if, for all* $x, y \in F$, *we have:*

(5) $|x + y| \leq \max(|x|, |y|)$.

The following example, known as the p-*adic valuation*, provides us with a nontrivially valued nonarchimedean field.

EXAMPLE 9.3. Let F be the field Q of all rational numbers and let p be any prime number. Every $x \in Q$, $x \neq 0$, can be written in a unique way in the form

$$x = p^n \cdot \frac{a}{b}$$

where a and b cannot be divided by p. We define the p-*adic valuation* of Q by setting

$$|x|_p = p^{-n}\ , \text{ if } x \neq 0;$$

$$|0|_p = 0.$$

Further examples of nonarchimedean valuations are provided by:

(a) the trivial valuation on any field;

(b) any valuation on a field with characteristic $p \neq 0$. In particular, all valuations of a finite field are nonarchimedean.

DEFINITION 9.4. *Let* $(F, |\cdot|)$ *be a valued field. If* $|\cdot|$ *is not nonarchimedean, we say that* $|\cdot|$ *is archimedean.*

Regarding archimedean valuations we have the following result. (See Monna [73]).

THEOREM 9.5. *(Ostrowski's Theorem) A field* F *with an archimedean valuation is isomorphic to a subfield of the field* $\mathbb{C}$ *of all complex numbers, and the valuation of* F *is then a power of the usual absolute value.*

Any valued field is a metric space. Indeed, for any x and y in a valued field $(F, |\cdot|)$, define the distance between them by

$$d(x, y) = |x - y|\ .$$

One easily verifies that d is a metric on F. Suppose now that $(F, |\cdot|)$ is nonarchimedean. Then

$$\begin{aligned} d(x, z) &= |x - z| = |(x - y) + (y - z)| \\ &\leq \max(|x - y|, |y - z|) \\ &= \max(d(x, y), d(y, z)). \end{aligned}$$

Thus, for all $x, y, z \in F$ we have

$$(6) \quad d(x,z) \leq \max(d(x,y), d(y,z)).$$

The above inequality is called the *ultrametric inequality*.

As a first example of what (6) way imply, consider on any valued field $(F, |\cdot|)$ the open ball (resp. closed ball) of radius r and center $x \in F$:

$$B_r(x) = \{y \in F; \ d(y, x) < r\}$$
$$= \{y \in F; \ |y - x| < r\}.$$
$$(\overline{B_r}(x) = \{y \in F; \ d(y, x) \leq r\}$$
$$= \{y \in F; \ |y - x| \leq r\}.)$$

Now, if the valuation is nonarchimedean, then for any $y_0 \in F$ in the closure of $B_r(x)$, these is a sequence $\{x_n\}$ in $B_r(x)$ such that $x_n \to y_0$ and therefore

$$|y_0 - x| = |y_0 - x_n + x_n - x|$$
$$\leq \max(d(y_0, x_n), \ d(x_n, x))$$
$$\leq d(x_n, x)$$

for $n \in \mathbf{N}$ sufficiently large, as $d(x_n, y_0) \to 0$. Now $d(x_n, x) < r$, and $y_0 \in B_r(x)$.

This shows that $B_r(x)$ is closed. Similarly, from (6) it follows that the set $\overline{B}_r(x)$ is open. Recalling that a topological space is said to be of *dimension* 0 if there is a basis of open sets formed by closed sets, we have proved:

PROPOSITION 9.6. *Every nonarchimedean valued field is of dimension* 0.

COROLLARY 9.7. *Every nonarchimedean valued field is totally disconnected.*

PROOF. Let x and y be distinct points in a nonarchimedean valued field $(F, |\cdot|)$. Let $r < d(x, y)$, $A = B_r(x)$ and

$$B = \{t \in F; \ t \notin B_r(x)\}.$$

Since $B_r(x)$ is closed, B is open. Therefore A and B are disjoint, non-empty, and open; moreover, $F = A \cup B$ with $x \in A$ and $y \in B$. This shows that F is totally disconnected.

THEOREM 9.8. *Let* X *be a Hausdorff space. Then* X *is of dimension* 0 *if, and only if,* $C_b(X;F)$ *is separating over* X, *for any nonarchimedean valued field* $(F, |\cdot|)$.

PROOF. Let X be a Hausdorff space of dimension 0, and let $(F, |\cdot|)$ be a nonarchimedean valued field. Let x and y be two distinct points in X. There exists a clopen set, i.e. an open set which is also closed, $U \subset X$, such that $x \in U$, $y \notin U$. The F-characteristic function ϕ_U defined by

$$\phi_U(t) = 1, \quad \text{if} \quad t \in U;$$

$$\phi_U(t) = 0, \quad \text{if} \quad t \notin U;$$

for all $t \in F$, is then an element of $C_b(X;F)$ such that

$$\phi_U(x) \neq \phi_U(y).$$

Conversely, assume $C_b(X;F)$ is separating over X. Let $x, y \in X$, with $x \neq y$. There exists $f \in C_b(X;F)$ with $f(x) \neq f(y)$.

Let $U = \{t \in X; |f(x) - f(t)| < |f(x) - f(y)|\}$. Then U is clopen, $x \in U$ and $y \notin U$.

§ 2. KAPLANSKY'S THEOREM.

If X is a compact Hausdorff space and $(F, |\cdot|)$ is a nonarchimedean valued field we endow $C(X;F)$ with the topology of uniform convergence on X, given by the sup-norm

$$f \to \|f\| = \sup\{|f(x)|;\ x \in X\}.$$

The following *separating version* of the Stone-Weierstrass theorem is due to I. Kaplansky [72].

THEOREM 9.9. *Let* $(F, |\cdot|)$ *be a nonarchimedean valued field, and let* X *be a compact Hausdorff space. Let* $A \subset C(X;F)$ *be a unitary subalgebra which is separating over* X . *Then* A *is uniformly dense in* C(X ; F).

PROOF. Let $f \in C(X;F)$. Since X is compact, $f(X) \subset F$ is compact. Now, for any $\varepsilon > 0$, f(X) is contained in the union of all open balls $B_\varepsilon(f(t))$, when $t \in X$, i.e.

$$f(X) \subset \bigcup \{B_\varepsilon(f(t));\ t \in X\}.$$

By compactness, there exists a finite set $\{t_1, t_2, \ldots, t_n\}$ such that

$$f(X) \subset B_\varepsilon(f(t_1)) \cup \ldots \cup B_\varepsilon(f(t_n)).$$

Let $B_1 = B_\varepsilon(f(t_1))$ and for $i = 2,3,\ldots,n$,

$$B_i = B_\varepsilon(f(t_i)) \setminus \bigcup_{j<i} B_\varepsilon(f(t_j)).$$

Then the sets B_i are a pairwise disjoint clopen cover of f(X). Therefore, the sets $U_i = f^{-1}(B_i)$, $i = 1,2,\ldots,n$, form a pairwise disjoint clopen cover of X. Let $t \in X$. Then $t \in U_i$ for some $i = 1,2,\ldots,n$ and $t \notin U_j$ for all $j \neq i$. Hence, if we put $\phi_k = \phi_{U_k}$, the characteristic function of U_k, $k = 1,2,\ldots,n$, then $\phi_i(t) = 1$ and $\phi_j(t) = 0$ for all $j \neq i$. Now $t \in U_i$ implies $|f(t) - f(t_i)| < \varepsilon$ and then $|f(t) - g(t)| < \varepsilon$, where

$$g = \sum_{k=1}^{n} f(t_k)\phi_k .$$

Therefore, all we have to prove is that g belongs to the closure of A in C(X ; F). To show this, it is sufficient to prove that the ϕ_k's belong to the closure of A, i.e. that if U is a clopen set in X, then ϕ_U belongs to the closure of A in C(X ; F). Let $0 < \varepsilon < 1$.

Fix $x \in X$, $x \notin U$. For each $t \in U$, since $x \neq t$, and A is a separating unitary subalgebra of C(X ; F), there is $g_t \in A$ such that $g_t(t) = 0$ and $g_t(x) = 1$. By Kaplansky's Lemma, there exists a polynomial $p_t : F \to F$, whose constant term is zero such that $p_t(1) = 1$ and $|p_t(s)| \leq 1$ for all s in the compact set $g_t(X) \subset F$. Then $f_t = p_t \circ g_t$ belongs to A,

$f_t(t) = 0$, $f_t(x) = 1$, and $\|f_t\| = \sup\{|f_t(y)|;\ y \in X\} \leq 1$. By continuity of f_t there exists an open neighborhood N_t of t such that $|f_t(y)| < \varepsilon$ for all $y \in N_t$. Since U is compact, there exists $t_1,\ldots,t_n \in U$ such that $U \subset N_{t_1} \cup \ldots \cup N_{t_n}$. Consider $f_x = f_{t_1} \cdot f_{t_2} \cdot \ldots \cdot f_{t_n}$. Then $f_x \in A$, $f_x(x)=1$, $\|f_x\| \leq 1$, and $|f_x(y)| < \varepsilon$ for all $y \in U$. Now $h_x = 1 - f_x$ belongs to A too, since $1 \in A$. Moreover, $h_x(x) = 0$ and $|1 - h_x(y)| < \varepsilon$ for all $y \in U$. By continuity of h_x there exists an open neighborhood W_x of x such that $|h_x(y)| < \varepsilon$ for all $y \in W_x$. Since $X \setminus U$ is compact, there exists $x_1, \ldots, x_m$ in $X \setminus U$ such that

$$X \setminus U \subset W_{x_i} \cup \ldots \cup W_{x_m}$$

Consider $h = h_{x_1} \cdot h_{x_2} \cdot \ldots \cdot h_{x_m}$. Then $h \in A$. Moreover, if $y \notin U$, then $y \in W_{x_i}$ for some $i=1,2,\ldots,m$, and $|h_{x_i}(y)| < \varepsilon$, while $\|h_{x_j}\| \leq 1$ for all $j = 1,2,\ldots,m$. Hence

(a) $\quad |h(y)| < \varepsilon$ for all $y \notin U$.

On the other hand, if $y \in U$, then $|1 - h_{x_i}(y)| < \varepsilon$ for all $i = 1,2,\ldots,m$. We claim that $|1 - h_{x_1}(y) \cdot \ldots \cdot h_{x_k}(y)| < \varepsilon$ for all $k = 1,2,\ldots,m$. This is clear for $k = 1$. Assume the claim for $1 < k = j < m$. Then

$$|1 - h_{x_1}(y) \cdot \ldots \cdot h_{x_{j+1}}(y)| =$$

$$= |1 - h_{x_{j+1}}(y) + h_{x_{j+1}}(y) - h_{x_1}(y) \cdot \ldots \cdot h_{x_j}(y)\, h_{x_{j+1}}(y)|$$

$$\leq \max(|1 - h_{x_{j+1}}(y)|, |h_{x_{j+1}}(y)| \cdot |1 - h_{x_1}(y) \cdot \ldots \cdot h_{x_j}(y)|)$$

$$\leq \max(|1 - h_{x_{j+1}}(y)|\ ,\ |1 - h_{x_1}(y) \cdot \ldots \cdot h_{x_j}(y)|\ )$$

because $|h_{x_{j+1}}(y)| \leq 1$. By the induction hypothesis, $|1 - h_{x_1}(y) \cdot \ldots \cdot h_{x_j}(y)| < \varepsilon$. Hence, $|1 - h_{x_1}(y) \cdot \ldots \cdot h_{x_{j+1}}(y)| < \varepsilon$, and the claim is true for j + 1. This proves the claim.

In particular, for $k = m$, we get

(b) $|1 - h(y)| < \varepsilon$, for all $y \in U$.

From (a) and (b), we see that $\|\phi_U - h\| < \varepsilon$, i.e. ϕ_U belongs to the closure of A in $C(X;F)$, for any clopen $U \subset X$, and this ends the proof.

REMARK. The Kaplansky's Lemma referred to in the above proof is the following.

LEMMA 9.10. *Let* $(F, |\cdot|)$ *be a nonarchimedean valued field and let* $x \in F$, $x \neq 0$. *Given any compact subset* $K \subset F$, *there exists a polynomial* $p : F \to F$ *whose constant term is zero such that* $p(x) = 1$ *and* $|p(t)| \leq 1$ *for all* $t \in K$.

For a proof see Lemma 1, Kaplansky [72]; or Narici, Beckenstein, and Bachman [76].

The first author to prove a Stone - Weierstrass Theorem for nonarchimedean valued fields was Dieudonné, who proved such a result in [70] for the field of p-adic numbers. Kaplansky's Theorem was extended to the general case of arbitrary Krull valuations, i.e. not necessarily real-valued valuations, by Chernoff, Rasala and Waterhouse. They proved the following.

THEOREM 9.11. *Let* F *be a field with an arbitrary Krull valuation, except* $\mathbb{C}$ *with its usual absolute value. Let* $A \subset C(X;F)$ *be a unitary subalgebra which is separating over* X. *Then* A *is uniformly dense in* $C(X;F)$.

For a proof see Chernoff, Rasala and Waterhouse [69]. Since we shall not treat the case of not necessarily real-valued valuations, we shall not use theorem 9.11 in the sequel.

Let us describe a quotient construction which permits to derive from Theorem 9.9, i.e. from the separating case, a *general version* of the Stone - Weierstrass Theorem, namely a version describing the closure of a unitary subalgebra of $C(X;F)$, whenever X is a compact Hausdorff space.

As in the case of $\mathbb{K}$-valued functions, given

$A \subset C(X;F)$, we denote by X/A the equivalence relation defined on X a follows: if $x, y \in X$, then we say that $x \equiv y$ (modulo X/A) if, and only if, $f(x) = f(y)$ for all $f \in A$. Let Y by the quotient topological space of X modulo X/A and let π be the quotient map of X onto Y; π is continuous and for each $x \in X$, $y = \pi(x)$ is the equivalence class $[x]$ of x modulo X/A.

Hence, for each $f \in A$, there is a unique $g : Y \to F$ such that $f(x) = g(\pi(x))$ for all x in X. We claim that g is continuous. Indeed, for every open subset $G \subset F$, the set $f^{-1}(G)$ is open in X, and $f^{-1}(G) = \pi^{-1}(g^{-1}(G))$. By the definition of the quotient topology of Y, this means that $g^{-1}(G)$ is an open subset of Y. Let us define $B \subset C(Y;F)$ by setting

$$B = \{g \in C(Y;F);\ f = g \circ \pi\ ,\ f \in A\}.$$

It follows that B is a subalgebra (resp. a unitary subalgebra) of $C(Y;F)$ whenever A is a subalgebra (resp. a unitary subalgebra) of $C(X;F)$. Notice the important fact that B is separating over Y. By Theorem 9.8, and the fact that $C_b(Y;F) = C(Y,F)$, as Y is compact, it follows that Y is of dimension 0.

THEOREM 9.12. *Let* X *be a compact Hausdorff space and let* $(F, |\cdot|)$ *be a nonarchimedean valued field. Let* $A \subset C(X;F)$ *be a unitary subalgebra, and let* $f \in C(X;F)$. *Then* f *belongs to the uniform closure of* A *in* $C(X;F)$ *if, and only if,* f *is constant on each equivalence class of* X *modulo* X/A.

PROOF. Clearly each $f \in C(X;F)$ which belongs to the uniform closure of A is constant on each equivalence class of X modulo X/A.

Conversely, let $f \in C(X;F)$ be constant on each equivalence class of X modulo X/A. Let Y, π and B as before. There exists $g : Y \to F$ such that $f = g \circ \pi$. As in the proof that B is contained in $C(Y;F)$, it is easy to see that g belongs to $C(Y;F)$. Now, since B is a separating unitary subalgebra of $C(Y;F)$, by Theorem 9.9, B is dense in $C(Y;F)$. Therefore g belongs to the uniform closure of B in $C(Y;F)$. Since the mapping $h \mapsto h \circ \pi$ is an isometry of $C(Y;F)$ into $C(X;F)$, it follows that f belongs to the uniform closure of A in C(X;F).

The hypothesis that the algebra A be unitary can be very annoying sometimes, so let us remove it.

THEOREM 9.13. *Let* X *and* F *be as in Theorem 9.12. Let* $A \subset C(X;F)$ *be a subalgebra, and let* $f \in C(X;F)$. *Then* f *belongs to the uniform closure of* A *in* C(X;F) *if, and only if, the following conditions hold:*

(1) *given* $x, y \in X$ *with* $f(x) \neq f(y)$, *there exists* $g \in A$ *such that* $g(x) \neq g(y)$;

(2) *given* $x \in X$ *with* $f(x) \neq 0$, *there exists* $g \in A$ *such that* $g(x) \neq 0$.

PROOF. Let $f \in C(X;F)$ be in the uniform closure of A. Let $x, y \in X$ with $f(x) \neq f(y)$. Let $\varepsilon = |f(x) - f(y)| > 0$. Assume that $g(x) = g(y)$ for $g \in A$. Let $g \in A$ be such that $\|f - g\| < \varepsilon$. Then

$$\varepsilon = |f(x) - f(y)| = |f(x) - g(x) + g(y) - f(y)|$$

$$\leq \max(|f(x) - g(x)|, |g(y) - f(y)|) < \varepsilon,$$

a contradiction. This proves (1). Analogously, one proves (2).

Conversely, let $f \in C(X;F)$ be a function satisfying conditions (1) and (2).

CASE I. There exists a point x in X such that $g(x) = 0$ for all functions g in A. By condition (2), we have $f(x) = 0$ too. Let $B \subset C(X;F)$ be the subalgebra generated by A and the constants. The equivalence relations X / A and X / B are the same, and by condition (1), f is then constant on each equivalence class of X modulo X / B. By Theorem 9.12, f belongs to the uniform closure of the unitary subalgebra $B \subset C(X;F)$. Let $\varepsilon > 0$ be given. There exists $g \in A$ and constant $\lambda \in F$ such that

$$|f(t) - g(t) - \lambda| < \varepsilon$$

for all $t \in X$. Making $t = x$, we obtain $|\lambda| < \varepsilon$. Since F is nonarchimedean,

$$|f(t) - g(t)| = |f(t) - g(t) - \lambda + \lambda| \leq$$

$$\leq \max(|f(t) - g(t) - \lambda|, |\lambda|) < \varepsilon,$$

for all $t \in X$. Hence, $\|f - g\| < \varepsilon$, and f belongs to the uniform closure of A.

CASE II. The algebra A has no common zeros. By Proposition 2, Chernoff, Rasala and Waterhouse [69], the algebra A contains a function h vanishing nowhere on X. Now $1/h$ belongs to $C(X;F)$ and is constant on each equivalence class modulo X/B. By Theorem 9.12, $1/h$ belongs to the uniform closure $\overline{B}$ of B in $C(X;F)$. Since A is a B-module, the uniform closure $\overline{A}$ of A in $C(X;F)$ is a $\overline{B}$-module. Hence $1 = h(1/h) \in \overline{A}$. This proves that $\overline{A}$ is a unitary subalgebra of $C(X;F)$. Since X/A and $X/\overline{A}$ are the same equivalence relation, by condition (1), f is constant on each equivalence class modulo $X/\overline{A}$. By Theorem 9.12, f belongs to the uniform closure of $\overline{A}$, namely $\overline{A}$ itself.

§ 3. NORMED SPACES.

Let $(F, |\cdot|)$ be a valued field, and let E be a vector space over F.

DEFINITION 9.14. *A mapping* $\|\cdot\| : E \to \mathbb{R}$ *is called a norm on* E *if*

(1) $\|x\| \geq 0$, *for all* $x \in E$;

(2) $\|x\| = 0$ *if, and only if* $x = 0$;

(3) $\|\lambda x\| = |\lambda| \cdot \|x\|$, *for all* $\lambda \in F$, $x \in E$;

(4) $\|x + y\| \leq \|x\| + \|y\|$, *for all* $x, y \in E$.

If $\|\cdot\|$ is a norm over E, we say that $(E, \|\cdot\|)$ is a *normed space* over the valued field $(F, |\cdot|)$.

DEFINITION 9.15. *Let* $(E, \|\cdot\|)$ *be a normed space over the valued field* $(F; |\cdot|)$. *We say that* $\|\cdot\|$ *is a nonarchimedean norm if, for all* $x, y \in E$, *we have*

$$(5) \qquad \|x + y\| \leq \max(\|x\|, \|y\|).$$

In this case, we say that $(E, \|\cdot\|)$ is a *nonarchimedean normed space* over the valued field $(F, |\cdot|)$.

CONVENTION 9.16. *From now on we shall assume that all normed spaces considered are not reduced to* $\{0\}$.

REMARKS 9.17.

(a) Let $(E, \|\cdot\|)$ be a nonarchimedean normed space. By convention 9.16, $E \neq \{0\}$. Hence there exists $x \in E$, witn $\|x\| > 0$. Therefore $(F, |\cdot|)$ is nonarchimedean too.

(b) Let F be any field, and let $|\cdot|$ be the trivial valuation of F. If E is any vector space over F, and we set $\|x\| = 1$ for all $x \in E$ with $x \neq 0$, and $\|0\| = 0$, then $\|\cdot\|$ is a nonarchimedean norm on E called the *trivial norm* on E.

§ 4. VECTOR - VALUED FUNCTIONS.

Let X be a compact Hausdorff space and let $(E, \|\cdot\|)$ be a normed space over a valued field $(F, |\cdot|)$. The vector space over F of all continuous E-valued functions on X, denoted by $C(X;E)$ is also a normed space over the valued field $(F, |\cdot|)$: just define

$$\|f\| = \sup\{ \|f(x)\| ; x \in X\}$$

for all $f \in C(X;E)$. When $(E, \|\cdot\|)$ is nonarchimendean, $(C(X;E), \|\cdot\|)$ is nonarchimedean too.

Let $A \subset C(X;F)$ be a subalgebra and let $W \subset C(X;E)$ be a vector subspace which is an A-module, i.e. $AW \subset W$. Our aim is to describe the closure of W in $C(X;E)$, or more generally given a function f in $C(X;E)$ to find the nonarchimedean distance of f from W, i.e. to find

$$d(f\,;W) = \inf\{\| f - g\|\,;\, g \in W\}.$$

To solve this problem in the line of argument of Chapter 1, we need a "partition of the unity" result. To this end, we shall adapt the proof of Rudin [55], section 2.13, to the nonarchimedean setting. Namely we shall prove the following.

LEMMA 9.18. *Let* Y *be a* 0-*dimensional compact Hausdorff space, and let* $V_1,\ldots,V_n$ *be a finite open covering of* Y. *Let* $(F,|\cdot|)$ *be a nonarchimedean valued field. There exists functions* $h_i \in C(Y;F)$, $i = 1,\ldots,n$, *such that*

(a) $h_i(y) = 0$ *for all* $y \notin V_i$, $i = 1,\ldots,n$;

(b) $\| h_i \| \leq 1$, $i = 1,\ldots,n$;

(c) $h_1 + \ldots + h_n = 1$ *on* Y.

PROOF. Each $y \in Y$ has a closed and open neighborhood $W(y) \subset V_i$ for some i (depending on y). By compactness of Y, there are points $y_1,\ldots,y_m$ such that $Y = W_1 \cup \ldots \cup W_m$, where we have set $W_j = W(y_j)$ for each $j = 1,\ldots,m$. If $1 \leq i \leq n$, let H_i be the union of those W_j which lie in V_i. Let $f_i \in C(Y;F)$ be the characteristic function of H_i, $i = 1,\ldots,n$. Define

$$h_1 = f_1$$
$$h_2 = (1 - f_1)\, f_2$$
$$\cdots\cdots\cdots$$
$$h_n = (1 - f_1)(1 - f_2)\ldots(1 - f_{n-1})\, f_n$$

Then $H_i \subset V_i$ implies that $f_i(y) = 0$ for all $y \notin V_i$ and so $h_i(y) = 0$ for $y \notin V_i$ too, $i = 1,\ldots,n$. This proves (a). Clearly $\|h_i\| \leq 1$, $i = 1,\ldots,n$, since h_i takes only the values 0 and 1, which proves (b). On the other hand, $Y = H_1 \cup \ldots \cup H_n$ and

$$h_1 + \ldots + h_n = 1 - (1 - f_1)(1 - f_2)\ldots(1 - f_n).$$

Hence, given $y \in Y$, at least one $f_i(y) = 1$ and therefore

$$h_1(y) + \dots + h_n(y) = 1.$$

This proves (c).

THEOREM 9.19. *Let* E *be a nonarchimedean normed space. Let* $A \subset C(X;F)$ *be a subalgebra and let* $W \subset C(X;E)$ *be a vector subspace which is an* A - *module. Let* $f \in C(X;E)$. *Then*

$$d(f,W) = \sup\{ d(f|S;\ W|S);\ S \in P_A \},$$

where P_A *denotes the set of all equivalence classes* $S \subset X$ *modulo* X / A.

Before proving Theorem 9.19 let us point out that it implies the following result.

THEOREM 9.20. *Let* E , A , W *and* f *be as in Theorem 9.19. Then* f *belongs to the uniform closure of* W *in* C(X;E) *if, and only if,* $f|S$ *is in the uniform closure of* $W|S$ *in* C(S;E) *for each equivalence class* $S \subset X$ *modulo* X / A.

The above Theorem 9.20 is the nonarchimedean analogue of Nachbin's Stone - Weierstrass Theorem for modules (Theorem 1.5) and 9.19 is the "strong" Stone - Weierstrass Theorem for modules (terminology of Buck [12]).

PROOF OF THEOREM 9.19. Let us put $d = d(f;W)$ and

$$c = \sup\{ d(f|S;\ W|S);\ S \in P_A\}.$$

Clearly, $c \leq d$. To prove the reverse inequality, let $\varepsilon < 0$. Without loss of generality we may assume that A is unitary. Indeed, the subalgebra A' of C(X;F) generated by A and the constants is unitary, and the equivalence relations X / A and X / A' are the same. Moreover, since W is a vector space, W is an A-module if, and only if, W is an A' - module.

Let Y be the quotient space of X modulo X / A, with quotient map π. For any $S \in P_A$, since $d(f|S;\ W|S) < c + \varepsilon$, there exists some function w_S in the A-module W such that $\| w_S(t) - f(t) \| < c + \varepsilon$ for all $t \in S$. Let

$$K_S = \{x \in X; \|w_S(x) - f(x)\| \geqslant c + \varepsilon\}.$$

Then K_S is compact and disjoint from S. Hence, for each $y \in Y$, $y \notin \pi(K_S)$, if $S = \pi^{-1}(y)$. This implies that

$$\bigcap \{\pi(K_S) \; ; \; S = \pi^{-1}(y), \; y \in Y\}$$

is empty. By the finite intersection property, there is a finite set $\{y_1, \ldots, y_n\} \subset Y$ such that $\pi(K_1) \cap \ldots \cap \pi(K_n) = \emptyset$, where $K_i = K_S$, for $S = \pi^{-1}(y_i)$, $i = 1,\ldots,n$. Let V_i be the open subset given by the complement of $\pi(K_i)$, $i = 1,\ldots,n$. Y is a 0-dimensional compact Hausdorff space. Hence, by Lemma 9.18 there exist functions $h_i \in C(Y;F)$, $i = 1,\ldots,n$, such that

(a) $h_i(y) = 0$ for all $y \notin V_i$, $i = 1,\ldots,n$;

(b) $\|h_i\| \leq 1$, $i = 1,\ldots,n$;

(c) $h_1 + \ldots + h_n = 1$.

Put $g_i = h_i \circ \pi$, so that we have $g_i \in C(X;F)$, $i = 1,\ldots,n$, and each g_i is constant on every equivalence class of X modulo X/A. By Theorem 9.12, g_i belongs to the closure of A in the space $C(X;F)$, for each $i = 1,2,\ldots,n$. Notice that $g_i(x) = 0$ for all $x \in K_i$, $i = 1,\ldots,n$, since $h_i(y) = 0$ for all $y \in \pi(K_i)$, $i=1,\ldots,n$. Moreover $\|g_i\| \leq 1$, $i = 1,\ldots,n$, and $g_1 + \ldots + g_n = 1$ on X. Let $g = \sum_{i=1}^{n} g_i w_i$ where $w_i = w_S$, with $S = \pi^{-1}(y_i)$, $i=1,\ldots,n$.

Then $\|g(x) - f(x)\| < c + \varepsilon$, for all $x \in X$. Indeed, for any $x \in X$ we have

$$\|g(x) - f(x)\| = \left\| \sum_{i=1}^{n} g_i(x)(w_i(x) - f(x)) \right\|$$

$$\leq \max_{1 \leq i \leq n} |g_i(x)| \cdot \|w_i(x) - f(x)\| .$$

Now, for each $1 \leq i \leq n$, either $x \in K_i$ and the $g_i(x) = 0$; or else $x \notin K_i$ and then

$$|g_i(x)| \cdot \| w_i(x) - f(x)\| \leq \|w_i(x) - f(x)\| < c + \varepsilon.$$

Let $M = \max\{\| w_i \| ;\ i = 1,\dots,n\}$ and choose $\delta > 0$ such that $\delta M < c + \varepsilon$. For each $i = 1,\dots,n$, there is $a_i \in A$ such that $\| a_i - g_i \| < \delta$. Let us define $w = \sum_{i=1}^{n} a_i w_i$. Then $w \in W$ and for all $x \in X$,

$$\|w(x) - g(x)\| < c + \varepsilon.$$

Indeed, for any $x \in X$ we have

$$\|w(x) - g(x)\| = \|\sum_{i=1}^{n} (a_i(x) - g_i(x))w_i(x)\|$$

$$\leq \max_{1\leq i\leq n} |a_i(x) - g_i(x)| \cdot \| w_i(x) \|$$

$$\leq \delta M < c + \varepsilon .$$

Finally, notice that $\|w(x) - f(x)\| = \|w(x) - g(x) + g(x) - f(x)\| \leq$ $\leq \max (\|w(x) - g(x)\| , \|g(x) - f(x)\|) < c + \varepsilon$, for all $x \in X$. Hence $d < c + \varepsilon$. Since $\varepsilon < 0$ was arbitrary, $d \leq c$.

THEOREM 9.21. *Let* E, A *and* W *be as in Theorem 9.19. For each* $f \in C(X;E)$, *there exists an equivalence class* $S \subset X$ *modulo* X/A *such that*

$$d(f;W) = d(f|S\ ;\ W|S).$$

PROOF. Let Y and π as before. For each $g \in W$, the fucntion

$$y \to \| f\,|\pi^{-1}(y) - g\,|\,\pi^{-1}(y)\|$$

is upper semicontinuous on Y, by Lemma 1, Machado and Prolla [39]. Hence

$$y \to \inf\{\| f\,|\pi^{-1}(y) - g\,|\,\pi^{-1}(y)\| \ ;\ g \in W\}$$

is upper semicontinuous on Y too, and therefore attains its supremum on Y. By Theorem 9.19, this supremum is d(f,W). Let then $y \in Y$ be the point where d(f;W) is attained and let $S = \pi^{-1}(y)$. Then

$$d(f;W) = \inf\{\| f|S - g|S\|, g \in W\} =$$

$$= d(f|S; W|S),$$

as desired.

COROLLARY 9.22. *Let* E , A *and* W *be as in Theorem 9.19. Assume that* A *is separating over* X. *Then* W *is dense in* C(X;E) *if and only if* $W(x) = \{g(x); g \in W\}$ *is dense in* E, *for each* $x \in X$. *More generally, for any* $f \in C(X;E)$, $f \in \overline{W}$ *if, and only if,* $f(x) \in \overline{W(x)}$ *in* E, *for each* $x \in X$.

REMARK 9.23. We saw in the proof of Theorem 9.19 that we may assume without loss of generality that A is unitary. Using this fact, it is clear that Theorem 9.13 was not nedeed here. On the other hand, Theorem 9.20 implies Theorem 9.13, without need for Proposition 2 of Chernoff, Rasala and Waterhouse [69]. Indeed, let $f \in C(X;F)$ be a function satisfying conditions (1) and (2) of Theorem 9.13. The arguments of Propositon 1.2, Chapter 1,are valid with $\mathbb{K}$ substituted for F. Hence, f|S is in the uniform closure of A|S in C(S;F) for each equivalence class $S \subset X$ modulo X/A. Since A is a module over itself, f belongs to the uniform closure of A in C(X;F), by Theorem 9.20.

Using Corollary 9.22 we can prove Kaplansky's result on ideals in function algebras. Let E be a nonarchimedean normed non-associative algebra with unit (= unitary) over a (necessarily) nonarchimedean field F; that is, E is a not necessarily associative linear algebra with unit e over F equipped with a nonarchimedean norm satisfying

(1) $\| uv \| \leq \| u \| \cdot \| v \|$ and

(2) $\| e \| = 1$.

Condition (1) implies that multiplication is jointly

continuous: If X is any compact Hausdorff space, $C(X;E)$ with pointwise operations and sup norm becomes a nonarchimedean normed algebra with unit too (over the same field F). Now the problem arises of characterizing the closed right (resp. left) ideals $I \subset C(X;E)$. Suppose that for every $x \in X$ a closed right (resp. left) ideal $I_x \subset E$ is given, and let us define

$$I = \{f \in C(X;E);\ f(x) \in I_x \text{ for all } x \text{ in } X\}.$$

Manifestly, I is a closed right (resp. left) ideal in $C(X;E)$. We shall prove that *any* closed right (resp. left) ideal in $C(X;E)$ has the above form. Namely we have the following.

THEOREM 9.24. (Kaplansky) *Let X be a 0-dimensional compact Hausdorff space. Let E be a nonarchimedean normed algebra with unit* e *over a (necessarily nonarchimedean) valued fielf F. Let $I \subset C(X;E)$ be a closed right (resp. left) ideal. For each $x \in X$, let I_x be the closure of $I(x)$ in E. Then I_x is a closed right (resp. left) ideal in E, and*

$$I = \{f \in C(X;E);\ f(x) \in I_x \text{ for all } x \text{ in } X\}.$$

PROOF: For every x in X, $I(x)$ is clearly a right (resp. left) ideal in E. Since the multiplication in E is jointly continuous, the closure I_x of $I(x)$ is a right (resp. left) ideal in E. We claim that I is a $C(X;F)$-module. Indeed, let $f \in I$ and $g \in C(X;F)$ be given. Define $h \in C(X;E)$ to be $x \to g(x)e$, where e is the unit of E. If I is a right ideal, then for all $x \in X$,

$$g(x)f(x) = g(x)\,[f(x)e] = f(x)\,[g(x)e] = f(x)h(x).$$

Since $fh \in I$, gf belongs to I. (The case of a left ideal is treated similarly.) It remains to apply Corollary 9.22 to the separating algebra $C(X;F)$ and the closed $C(X;F)$-module I.

COROLLARY 9.25. *Under the hypothesis of Theorem 9.24 assume, that the algebra E is simple. Then any two-sided closed ideal consists of all functions vanishing on a closed subset of X,* *and the ideal is maximal if and only if the closed set is a singleton.*

PROOF: We first recall that the unitary algebra E is said to be *simple* if it has no two-sided ideals other than 0 and E. Let $N \subset X$ be a closed subset of X. Clearly, the subset

$$Z(N) = \{f \in C(X;E);\ f(x) = 0 \quad \text{for all } x \text{ in } N\}$$

is a closed two-sided ideal of $C(X;E)$.

Conversely, if I is a closed two-sided ideal in $C(X;E)$, let us define $N = \{x \in X;\ f(x) = 0 \quad \text{for all } f \in I\}$. Clearly, N is closed in X and $I \subset Z(N)$. Conversely, let $f \in Z(N)$, and assume by contradiction that $f \notin I$. By Theorem 9.24, there is some $x \in X$ such that $f(x) \notin I_x$. Therefore, $f(x) \neq 0$. Now $f \in Z(N)$, so $x \notin N$. However, I_x is a two-sided ideal in the simple algebra E, so $I_x = \{0\}$. Now $I_x = \{0\}$ implies $I(x) = 0$, and so $x \in N$. This contradiction shows that $f \in I$.

Let $L = C(X;E), x \in X$. Let $M = \{f \in L,\ f(x) = 0\}$. Then M is a closed two-sided ideal in L. Clearly, $M(x) = 0$. Let $y \neq x$. Let $f \in C(X;F)$ be such that $f(x) \neq f(y)$. Such an f exists because X is 0-dimensional. In fact we may assume $f(x) = 0$ and $f(y) = 1$. Therefore $f \in M$. Let $a \in E$. Then $af \in M$ and $af(y) = a$. Thus $M(y) = E$, for all $x \neq y$. Let $I \subset C(X;E)$ be a two-sided ideal containing M. Then $I(y) = E$ for all $y \neq x$. Now $I(x)$ is a two-sided ideal in E. Since E is simple, either $I(x) = 0$ or $I(x) = E$. By Theorem 9.24, this means that either $I = M$ or $I = C(X;E)$. Therefore M is maximal.

Conversely, if **M** is a maximal proper closed two-sided ideal in $C(X;E)$, then by Theorem 9.24, $\overline{M(x)} \neq E$ for some $x \in X$. Since E is simple, $M(x) = 0$. Therefore

$$M \subset \{f \in C(X;E);\ f(x) = 0\}.$$

Since the right-hand side of this inclusion is a proper closed ideal and M is maximal, $M = \{f \in C(X;E);\ f(x) = 0\}$.

§ 5. VECTOR FIBRATIONS.

As in Chapter 1 one can extend the results of the preceding §4 to the case of vector fibrations.

By a *vector fibration* we mean a pair $(X; (E_x; x \in X))$, where X is a Hausdorff topological space and $(E_x ; x \in X)$ is a family of vector spaces over the same scalar field F. The product set $\prod_{x \in X} E_x$ is made into a vector space over F in the usual way and its elements are called the *cross - sections* of the given vector fibration.

A vector space W of cross - sections, i. e. a vector subspace $W \subset \prod_{x \in X} E_x$, is said to be a *module over a subalgebra* $A \subset C(X;F)$, or an A - *module*, if for any $f \in A$ and $g = (g(x); x \in X)$ in W, we have $fg = (f(x)g(x); x \in X) \in W$.

We shall restrict our attention to vector spaces $L \subset \prod_{x \in X} E_x$ of cross - sections satisfying the following conditions:

(1) X is a compact Hausdorff space;

(2) each E_x is a nonarchimedean normed space over the same valued field $(F, |\cdot|)$; the norm in each E_x will be denoted by $t \mapsto \|t\|$;

(3) for each $f \in L$ the function $x \mapsto \|f(x)\|$ is upper semicontinuous on the space X.

From (1) - (3) it follows that L is a nonarchimedean normed space over $(F; |\cdot|)$, if we define

$$\|f\| = \sup\{\|f(x)\|; x \in X\}$$

for all $f \in L$. All the results of §4 can be extended to such *nonarchimedean normed spaces of cross - sections*, and the proofs are exactly the same. In particular, Theorem 9.19 reads as follows.

THEOREM 9.26. *Let* L *be a nonarchimedean normed space of cross-sections satisfying conditions* (1) - (3) *above. Assume that* L *is an* A - *module, where* A *is a subalgebra of* $C(X;F)$. *For every*

A - *submodule* $W \subset L$; *we have*

$$d(f;W) = \sup \{d(f|S;\ W|S);\ S \in P_A\},$$

for all $f \in L$, *where* P_A *denotes the set of all equivalence classes* $S \subset X$ *modulo* X/A.

COROLLARY 9.27. *Let* L, A *and* W *be as in Theorem 9.26. For each* $f \in L$, *there exists an equivalence class* $S \subset X$ *modulo* X/A *such that*

$$d(f;W) = d(f|S;\ W|S).$$

PROOF: Apply Lemma 1, Machado and Prolla [39].

COROLLARY 9.28. *Let* L, A *and* W *be as in Theorem 9.26. Assume that* A *is separating over* X. *Then* W *is dense in* L *if, and only if* $W(x) = \{g(x);\ g \in W\}$ *is dense in* $L(x) \subset E_x$ *for each* $x \in X$. *More generally, for any* $f \in L$, f *belongs to the closure of* W *in* L *if and only if* $f(x)$ *belongs to the closure of* $W(x)$ *in* $L(x)$ *for each* $x \in X$.

To state the analogous result of Theorem 9.24 for vector fibrations, assume that besides (1) - (3) the following further conditions are satisfied:

(4) each E_x is a not necessarily associative linear algebra with unit e_x over the same nonarchimedean valued field $(F, |\cdot|)$; moreover each E_x is equipped with a nonarchimedean norm satisfying

(4.1) $\|uv\| \leq \|u\| \cdot \|v\|$, for all $u, v \in E_x$

(4.2) $\|e_x\| = 1$

(5) $e = (e_x;\ x \in X)$ belongs to L.

DEFINITION 9.29. *If* (1) - (5) *are satisfied we say that* L *is a nonarchimedean normed unitary algebra of cross - sections. Indeed, if we define operations coordinatewise in* $\prod_{x \in X} E_x$, *this becomes a not necessarily associative linear algebra over* F, *with unit* $e = (e_x ;\ x \in X)$, L *is a subalgebra containing the unit* e, *and the norm of* L *satisfies:*

(a) $\| fg \| \leq \| f \| \cdot \| g \|$ *for all* $f, g \in L$

(b) $\| e \| = 1$.

If L is *complete* with respect to the metric induced by the norm of L we say that L is a *nonarchimedean unitary Banach algebra of cross - sections.*

DEFINITION 9.30. *Suppose that* L *is a unitary nonarchimedean normed algebra of cross - sections over* X. *Let* $W \subset L$ *be a vector sub — space. Following Murphy* [74], *one says that a cross - section* $f \in L$ *is in* W *locally at a point* $x \in X$, *if for each* $\varepsilon > 0$ *there is a neighborhood* U *of* x *in* X *and an element* $g \in W$ *such that*

$$\| f(y) - g(y) \| < \varepsilon$$

for all $y \in U$. *The vector subspace* W *is said to be local if the uniform closure* $\overline{W}$ *of* W *in* L *contains all the cross - sections which are in* W *locally at all points of* X. *Let* $\tilde{\Delta}(W) = \{f \in L;\ f$ *is in* W *locally at all points of* $X\}$. *Clearly,* $\tilde{\Delta}(W) \supset \overline{W}$, *and* W *is local if and only if* $\tilde{\Delta}(W) \subset \overline{W}$.

DEFINITION 9.31. *Let* L *be a unitary nonarchimedean normed algebra of cross - sections over* X. *We say that* L *is separating if for all pairs of distinct points* $x, y \in X$ *there is* $f \in L$ *such that* $\| f \| \leq 1$, $f(x) = 0$, *and* $f(y) = e_y$.

Let $U \subset X$ be any subset of X. The *characteristic cross - section of* U, denoted by ϕ_U is defined by

$$\phi_U(x) = e_x \quad \text{if} \quad x \in U$$

$$\phi_U(x) = 0 \quad \text{if} \quad x \notin U.$$

DEFINITION 9.32. *Let* L *be as in Definition 9.31. We say that* L *is full if* L *contains the characteristic cross - sections of all clopen subsets of* X .

If X is 0 - dimensional, then every local unitary subalgebra of a full algebra is separating. The converse is Murphy's Stone - Weierstrass Theorem (see [74]). We state if for A - modules $W \subset L$. Recall that a vector subspace $W \subset L$ is called a *right* (resp. *left*) *module* over an algebra $A \subset L$ if $AW \subset W$ (resp. $WA \subset W$). If $W \subset L$ is a right module or a left module over A , we say simply that W is an A - *module*. When $W \subset L$ is both a right and a left module over A , we say that W is an A-*bimodule*. Clearly, a subalgebra $A \subset L$ is a bimodule over itself.

THEOREM 9.33. *Let* L *be a full unitary nonarchimedean normed algebra of cross - sections over a* 0 - *dimensional compact Hausdorff space* X . *Let* $A \subset L$ *be a separating unitary subalgebra, and let* $W \subset L$ *be a vector subspace which is an* A - *module. Then* W *is local.*

PROOF: Assume W is a right module over A . Let $f \in \tilde{\Delta}(W)$ and $0 < \varepsilon < 1$ be given. For each $x \in X$, there is some $g_x \in W$ and some clopen neighborhood of x , say U_x such that

$$\|f(y) - g_x(y)\| < \varepsilon ,$$

for all $y \in U_x$. By compactness of X , there is a finite set $\{x_1, x_2, \ldots, x_n\} \subset X$ such that

$$X \subset U_{x_1} \cup \ldots \cup U_{x_n}$$

Let $U_1 = U_{x_1}$, and for $i = 2,3,\ldots,n$

$$U_i = U_{x_i} - \bigcup_{j<i} U_{x_j}$$

Then, the U_i's form a pairwise disjoint clopen cover of X . Let ϕ_i be the characterist cross - section of U_i , $i = 1,2,\ldots,n$.

Assume that we have proved that the uniform closure $\bar{A}$

of A is full. Then $\phi_i \in \overline{A}$ for all $i = 1,2,\ldots,n$. Hence $\phi_i g_{x_i}$ belongs to $\overline{A}W \subset \overline{A} \cdot \overline{W} \subset \overline{AW} \subset \overline{W}$, for all $i = 1,2,\ldots,n$. Hence g defined by

$$g = \phi_1 g_{x_1} + \ldots + \phi_n g_{x_n}$$

belongs to $\overline{W}$, and given $y \in X$,

$$\| f(y) - g(y) \| = \| f(y) - g_{x_i}(y) \| < \varepsilon$$

where $i = 1,2,\ldots,n$ is the unique index i such that $y \in U_i$. Therefore $\| f - g \| < \varepsilon$ and this shows that $f \in \overline{W}$, i.e. W is local.

It remains to prove that $\overline{A}$ is full. Let ϕ_U be the characteristic cross-section of a clopen subset $U \subset X$. We claim that $\phi_U \in \overline{A}$. Let $0 < \varepsilon < 1$ be given. Let $x \notin U$ and $y \in U$ be given. Since A is separating, there is $g_y \in A$ such that

$$\| g_y \| \leq 1, \; g_y(x) = e_x , \quad \text{and} \quad g_y(y) = 0.$$

Since $t \mapsto \| g_y(t) \|$ is upper semicontinuous there is a clopen neighborhood V_y of y in X such that

$$\| g_y(t) \| < \varepsilon,$$

for all $t \in V_y$. Now U is contained in the union of all V_y's, $y \in U$. By compactness, there is a finite set $\{y_1, y_2, \ldots, y_n\} \subset U$ such that

$$U \subset V_{y_1} \cup \ldots \cup V_{y_n} .$$

Let $g_x = g_{y_1} \cdot g_{y_2} \cdot \ldots \cdot g_{y_n} \in A$. Moreover

$$\| g_x \| \leq 1 , \quad g_x(x) = e_x , \quad \text{and}$$

$$\| g_x(t) \| < \varepsilon , \quad \text{for all} \quad t \in U,$$

because $0 < \varepsilon < 1$. Let $h_x = e - g_x$. Then $h_x \in A$, because A is unitary, and

$$\|h_x\| \leq 1, \ h_x(x) = 0.$$

For all $t \in U$, we have

$$\|e_t - h_x(t)\| = \|g_x(t)\| < \varepsilon .$$

By upper semicontinuity of h_x there is a clopen neighborhood W_x of x in X such that $\|h_x(t)\| < \varepsilon$ for all $t \in W_x$.

Now $X \setminus U$ is contained is the union of all W_x's, $x \in X \setminus U$. By compactness, there is a finite set

$$\{x_1, x_2, \dots, x_m\} \subset X \setminus U \quad \text{such that}$$

$$X \setminus U \subset W_{x_1} \cup \dots \cup W_{x_m} .$$

Let $h = h_{x_1} \cdot h_{x_2} \cdot \dots \cdot h_{x_m} \in A$. Moreover

$$\|h\| \leq 1, \quad \text{and for all} \quad t \in X \setminus U$$

$$\|h(t)\| < \varepsilon .$$

We claim that $\|\phi_U - h\| < \varepsilon$. Indeed, if $t \notin U$, then $\|\phi_U(t) - h(t)\| = \|h(t)\| < \varepsilon$. If $t \in U$, then for each $i = 1,2,\dots,m$, we have

$$\|e_t - h_{x_i}(t)\| < \varepsilon, \ \|h_{x_i}\| \leq 1. \quad \text{So}$$

$$\|e_t - h_{x_1}(t) \cdot \dots \cdot h_{x_k}(t)\| < \varepsilon \quad \text{for all } k = 1,2,\dots,m.$$

Indeed, this is trivial for $k = 1$. Assume it is true for some $1 < k = j < m$. Let $v_s = h_{x_s}(t)$ for any $s = 1,2,\dots,j+1$.Then we have

$$\|e_t - v_1 v_2 \dots v_j v_{j+1}\| =$$

$$= \|e_t - v_{j+1} + v_{j+1} - v_1 v_2 \cdots v_j v_{j+1}\|$$

$$\leq \max(\|e_t - v_{j+1}\|, \|v_{j+1}\| \cdot \|e_t - v_1 v_2 \cdots v_j\|)$$

$$< \varepsilon.$$

This proves our claim by induction. In particular, for $k = m$, we get

$$\|\phi_U(t) - h(t)\| = \|e_t - h(t)\| < \varepsilon.$$

This ends the proof that $\overline{A}$ is full.

COROLLARY 9.34. *Assume the hypothesis of Theorem 9.33. Then*

$$\overline{W} = \{f \in L;\ f(x) \in \overline{W(x)} \ \textit{for all} \ x \in X\}.$$

PROOF: Clearly one has the inclusion

$$\overline{W} \subset \{f \in L;\ f(x) \in \overline{W(x)} \ \text{for all} \ x \in X\}.$$

Conversely, let $f \in L$ be such that $f(x) \in \overline{W(x)}$ for all $x \in X$. We have to prove that $f \in \overline{W}$. By Theorem 9.33, W is local, so it is sufficient to prove that $f \in \tilde{\Delta}(W)$. Now, if $x \in X$ and $\varepsilon > 0$ are given, there is $g \in W$ such that $\|g(x) - f(x)\| < \varepsilon$. By upper semicontinuity this is true in a neighborhood U of x in X, i.e. $\|g(t) - f(t)\| < \varepsilon$ for all $t \in U$. Hence $f \in \tilde{\Delta}(W)$.

THEOREM 9.35. *(Vector-valued Kaplansky's Theorem). Let* X *be a* 0-*dimensional compact Hausdorff space, and let* E *be a nonarchimedean normed unitary algebra over a valued field* $(F, |\cdot|)$. *Let* $A \subset C(X;E)$ *be a separating unitary subalgebra* $A \subset C(X;E)$, *and let* $W \subset C(X;E)$ *be an* A-*module. Then*

$$\overline{W} = \{f \in C(X;E);\ f(x) \in \overline{W(x)} \ \textit{for all} \ x \in X\}.$$

PROOF: Consider the vector-fibration given by $E_x = E$ for all $x \in X$ and take $L = C(X;E)$. Then, for any clopen subset $U \subset X$, the characteristic vector valued function $\phi_U : X \to E$ is continuous. Hence $C(X;E)$ is full. It remains to apply Corollary 9.34.

COROLLARY 9.36. *Let* X *and* E *be as in Theorem 9.35. Let* $A \subset C(X;E)$ *be a separating unitary subalgebra. Then*

$$\bar{A} = \{ f \in C(X;E);\ f(x) \in \overline{A(x)} \quad \text{for all} \quad x \in X\} .$$

PROOF: Consider $W = A$ in Theorem 9.35, and recall that any algebra is a bimodule over itself.

COROLLARY 9.37. *Let* X *and* E *be as in Theorem 9.35. Let* $A \subset C(X;E)$ *be a separating subalgebra containing the constants. Then* A *is uniformly dense in* $C(X;E)$.

PROOF: Since A contains the constants, A is unitary and $A(x) = E$ for all $x \in X$. It remains to apply Corollary 9.36.

THEOREM 9.38. *(Kaplansky's Theorem) Let* X *be a* 0*-dimensional compact Hausdorff space and let* $(F, |\cdot|)$ *be a nonarchimedean valued field. Let* $A \subset C(X;F)$ *be a unitary subalgebra which separates the points of* X. *Then* A *is uniformly dense in* $C(X;F)$.

PROOF: By Corollary 9.37, taking $E = F$, it is enough to prove that A is separating in the sense of Definition 9.31. Let then $x \neq y$ be given in X. By hypothesis there in an element $f \in A$ such that $f(x) \neq f(y)$. Since F is a field, and A contains the constants, it follows that there is $g \in A$ such that $g(x) = 0$ and $g(y) = 1$. Now $g(X) \subset F$ is a compact subset, and therefore by Kaplansky's Lemma there is a polynomial $p : F \to F$ such that $p(0) = 0$, $p(1) = 1$, and $|p(t)| \leq 1$ for all $t \in g(X)$. Then $h = p \circ g$ belongs to A, $\| h \| \leq 1$, $h(x) = 0$, and $h(y) = 1$. Therefore A is separating in the sense of Definition 9.31.

REMARK 9.39. The remarks preceding Theorem 9.33 and the arguments in the proof of Theorem 9.33 show that, if X and L are as in Theorem 9.33, then for any *closed* unitary subalgebra $A \subset L$

the following are equivalent

(i) A is separating.

(ii) A is full.

(iii) A is local.

Let us now consider the characterization of closed ideals. For the case of $E_x = E$ for all $x \in X$, and continuous functions, this was done in Theorem 9.24 and Corollary 9.25 above.

THEOREM 9.40. *Let* L *be as in Theorem 9.33. Assume that* L *is essential. Let* $I \subset L$ *be a closed right (resp. left) ideal. For each* $x \in X$, *let* I_x *be the closure of* $I(x)$ *in* E_x. *Then* I_x *is a closed right (resp. left) ideal in* E_x, *and*

$$I = \{f \in L;\ f(x) \in I_x \quad \textit{for all} \quad x \quad \textit{in} \quad X\}.$$

PROOF: Take $A = L$ and $W = I$ in Corollary 9.34, and notice that by Remark 9.39, L full implies that L is separating. Hence

$$I = \{f \in L;\ f(x) \in I_x \quad \text{for all} \quad x \in X\}.$$

Now it is easy to show that $I(x)$ is a right ideal in $L(x)$. Indeed, if $a \in L(x)$, there is $f \in L$ such that $f(x) = a$. If $b \in I(x)$, there is $g \in I$ such that $g(x) = b$. Now $gf \in I$ and $ba = g(x)\, f(x) = (gf)(x)$ belongs to $I(x)$. Thus $I(x)$ is a right ideal in $L(x)$. The hypothesis that L is essential means that $L(x) = E_x$ for all $x \in X$, and this ends the proof of Theorem 9.40.

COROLLARY 9.41. *Let* L *be as in Theorem 9.40, and assume that for each* $x \in X$, E_x *is simple. Then any closed two-sided ideal consists of all functions vanishing on a closed subset of* X. *Moreover, every maximal closed two-sided proper ideal is of the form* $\{f \in L;\ f(x) = 0\}$ *for some* $x \in X$.

PROOF: Let N be a closed subset of X. It is easy to see that

$$Z(N) = \{f \in L;\ f(x) = 0 \quad \text{for all} \quad x \in N\}$$

is a closed two-sided ideal of L.

Conversely, let $I \subset L$ be a closed two-sided ideal in L. Define

$$N(I) = \{x \in X;\ f(x) = 0 \quad \text{for all} \quad f \in I\}.$$

It is easy to see that N is closed in X and clearly $I \subset Z(N)$. Conversely, let $f \in Z(N)$, and assume by contradiction that $f \notin I$. By Theorem 9.40, there is some $x \in X$ such that $f(x) \notin I_x$. Hence $f(x) \neq 0$. Now $f \in Z(N)$, so $x \notin N$. On the other hand, I_x is a two-sided closed ideal in the simple algebra E_x. Since $f(x) \notin I_x$, $I_x \neq E_x$. Therefore $I_x = \{0\}$, and this implies $I(x) = 0$. Hence $x \in N$. This contradiction shows that $f \in I$.

For $x \in X$, set $M = \{f \in L;\ f(x) = 0\}$. M is a closed two-sided ideal in L. Clearly, $M(x) = 0$. Let $y \in X$ be a distinct point. Since X is 0-dimensional, there is a clopen neighborhood U of y, with $x \notin U$. Since L is full, $\phi_U \in L$. If $a \in E_y = L(y)$ (because L is essential), there is some $f \in L$ with $f(y) = a$. Hence $g = \phi_U f$ is such that

$$g(x) = 0f(x) = 0, \quad g(y) = e_y\, a = a.$$

Hence $g \in M$ and $g(y) = a$. Therefore $M(y) = E_y$, for all $y \neq x$. Let then $I \subset L$ be a two-sided ideal containing M. Therefore $I(y) = E_y$ for all $y \neq x$. Now $I(x)$ is a two-sided ideal in E_y. Since E_y is simple either $I(x) = 0$ or $I(x) = E_x$. In the first case, we have

$$\overline{I(t)} = \overline{M(t)}, \quad \text{for all} \quad t \in X.$$

By Corollary 9.34, $I = M$.

In the second case, we have

$$\overline{I(t)} = \overline{L(t)}, \quad \text{for all} \quad t \in X.$$

By the same Corollary, $I = L$. Therefore M is maximal.

Conversely if M is maximal closed two-sided proper ideal in L, then $\overline{M(x)} \neq E_x$ for some $x \in X$. Since E_x is simple, $M(x) = 0$. Therefore $M \subset \{f \in L;\ f(x) = 0\}$. Since $\{f \in L;\ f(x) = 0\}$ is proper and M is maximal, we have

$$M = \{f \in L;\ f(x) = 0\}.$$

This ends the proof of Corollary 9.41.

§ 6. SOME APPLICATIONS.

In this section X is a compact Hausdorff space and E is a nonarchimedean normed space over a valued field $(F, |\cdot|)$. By convention 9.16, $E \neq \{0\}$. The vector subspace of $C(X;E)$ consisting of all finite sums of functions of the form $x \to f(x)v$, where $f \in C(X;F)$ and $v \in E$, will be denoted by $C(X;F) \otimes E$. Clearly, $C(X;F) \otimes E$ is a $C(X;F)$-module.

THEOREM 9.42. *Let* X *be a 0-dimensional compact Hausdorff space. Then* $C(X;F) \otimes E$ *is uniformly dense in* $C(X;E)$.

PROOF: Let $W = C(X;F) \otimes E$. Then W is a $C(X;F)$-module, and $C(X;F)$ is separating over X. For each $x \in X$, $W(x) = E$. By Corollary 9.22, W is dense in $C(X;E)$.

If X and Y are two compact Hausdorff spaces, $C(X;F) \otimes C(Y;F)$ denotes the vector subspace of $C(X \times Y;F)$ consisting of all finite sums of functions of the form

$$(x, y) \to f(x)g(y)$$

where $f \in C(X;F)$ and $g \in C(Y;F)$. If both X and Y are 0-dimensional spaces, then $C(X;F) \otimes C(Y;F)$ is a separating unitary subalgebra of $C(X \times Y;F)$.

THEOREM 9.43. *Let* X *and* Y *be two 0-dimensional compact Hausdorff spaces. Then* $(C(X;F) \otimes C(Y;F)) \otimes E$ *is uniformly dense in* $C(X \times Y;E)$.

PROOF: Let $W = (C(X;F) \otimes C(Y;F)) \otimes E$. W is $C(X;F) \otimes C(Y;F)$ - module such that $W(x,y) = E$ for every pair $(x,y) \in X \times Y$. The result now follows from Corollary 9.22.

REMARK: When $E = F$, then the space $(C(X;F) \otimes C(Y;F)) \otimes E$ is just $C(X;F) \otimes C(Y;F)$ and one obtains Dieudonné's Théorème 2, [70] .

In Chapter 4 we studied polynomial algebras of functions with values in vector spaces over $\mathbb{R}$ or $\mathbb{C}$. To study the nonarchimedean analogue let us adopt the following

DEFINITION 9.44. *A vector subspace* $W \subset C(X;E)$ *is called a polynomial algebra if* $A = \{u(f);\ u \in E',\ f \in W\}$ *is a subalgebra of* $C(X;F)$ *such that* $A \otimes E \subset W$.

Let us give an example of a polynomial algebra. Let

$$P_f(E;F) \subset C(E;F)$$

be the algebra over F generated by the topological dual E' of E. An element $p \in P_f(E;F)$ is called a *continuous polynomial of finite type from* E *into* F, and is of the from

$$(1) \qquad p = \sum_{|\kappa| \leq m} a_\kappa u^\kappa$$

where $\kappa = (\kappa_1,\ldots,\kappa_n) \in \mathbb{N}^n$, $n \in \mathbb{N}^*$, $|\kappa| = \kappa_1+\ldots+\kappa_n$, $m \in \mathbb{N}$, $a_\kappa \in F$, $u = (u_1,\ldots,u_n) \in (E')^n$, and we define

$$(2) \qquad u^\kappa(t) = (u_1(t))^{\kappa_1} \ldots (u_n(t))^{\kappa_n}$$

for all $t \in E$. Let us now consider two nonarchimedean normed spaces E_1 and E_2 over the same nonarchimedean valued field F. We define $P_f(E_1, E_2)$ as the vector subspace of $C(E_1, E_2)$ generated by the functions of the form $t \in E_1 \to p(t)v$ where $p \in P_f(E_1, F)$ and $v \in E$. Let now $A = \{u(p);\ u \in E_2';\ p \in P_f(E_1,E_2)\}$. Clearly, $A \subset P_f(E_1;F)$, and $A \otimes E_2 \subset P_f(E_1, E_2)$. Suppose $(E_2)' \neq \{0\}$. Then $A = P_f(E_1;F)$ and $P_f(E_1;E_2)$ is a polynomial algebra. Also, if $X \subset E_1$ is any compact subset, then

$W = P_f(E_1 ; E_2)|X$ is a polynomial algebra contained in $C(X;E_2)$. More generally, if $S \subset C(X;F)$ is any subset, let $A \subset C(X;F)$ be the subalgebra over F generated by S. If $E' \neq \{0\}$, then $W = A \otimes E$ is a polynomial algebra. Indeed, in this case we have $A = \{u(f);\ u \in E', f \in W\}$. In particular, $C(X;E)$ is a polynomial algebra, when $E' \neq 0$ (e. g., when $E = F$).

When the field F is *spherically complete*, the Hahn - Banach Theorem is valid for any nonarchimedean normed space E over F (see Ingleton [71]), and then E' is separating over E, and a fortiori, $E' \neq \{0\}$.

Let us introduce the following notation. If $W \subset C(X;E)$ is an A-module, where $A \subset C(X;F)$, we denote by $L_A(W)$ the set of all $f \in C(X;E)$ such that the restriction $f|S$ in the uniform closure of $W|S$ in $C(S;E)$, for each equivalence class $S \subset X$ modulo X/A. Thus, if $\overline{W}$ denotes the uniform closure of W in $C(X;E)$, the Theorem 9.20 may be stated as $f \in \overline{W} \Longleftrightarrow f \in L_A(W)$.

THEOREM 9.45. *Let* E *be a nonarchimedean normed space such that* E' *is separating over* E, *and let* $W \subset C(X;E)$ *be a polynomial algebra. Let* $A = \{u(g);\ u \in E', g \in W\}$. *Then, for every* $f \in C(X;E)$ *the following conditions are equivalent.*

(1) $f \in \overline{W}$;

(2) *given* $x, y \in X$ *and* $\varepsilon > 0$, *there is* $g \in W$ *such that* $\|f(x) - g(x)\| < \varepsilon$ *and* $\|f(y) - g(y)\| < \varepsilon$;

(3) (a) *given* $x, y \in X$, *with* $f(x) \neq f(y)$, *there is* $g \in W$ *such that* $g(x) \neq g(y)$; *and*

(b) *given* $x \in X$, *with* $f(x) \neq 0$, *there is* $g \in W$ *such that* $g(x) \neq 0$;

(4) $f \in L_A(A \otimes E)$.

PROOF: (1) $\Rightarrow$ (2). Obvious.

(2) $\Rightarrow$ (3). Let $x, y \in X$ with $f(x) \neq f(y)$. Define $\varepsilon = \|f(x) - f(y)\| > 0$. By (2) there is $g \in W$ such that

$$\|f(x) - g(x)\| < \varepsilon \quad \text{and} \quad \|f(y) - g(y)\| < \varepsilon.$$

If $g(x) = g(y)$, then $\varepsilon = \|f(x) - g(x) + g(y) - f(y)\| \leq$

$$\max(\|f(x) - g(x)\|, \|g(y) - f(y)\|) < \varepsilon,$$

a contradiction. This proves (a). The proof of (b) is similar.

(3) $\Longrightarrow$ (4). Let $S \subset X$ be an equivalence class modulo X/A, and let $x, y \in S$. If $f(x) \neq f(y)$, by (a) there is $g \in W$ such that $g(x) \neq g(y)$. Since E' is separating over E, there is $u \in E'$ such that $u(g(x)) \neq u(g(y))$. This is impossible, because $u(g) \in A$. Hence f is constant over S. Let $t \in E$ be its constant value. If $t = 0$, then $0 \in A \otimes E$ agrees with f over S. If $t \neq 0$, then, by (b) there is $g \in W$ such that $g(x) \neq 0$, where $x \in S$ is chosen arbitrarily. Let now $u \in E'$ be such that $u(g(x)) = 1$. Then the function

$$h = u(g) \otimes t$$

belongs to $A \otimes E$ and agrees with f over S. Therefore

$$f \in L_A(A \otimes E).$$

(4) $\Longrightarrow$ (1). By Theorem 9.20 applied to the A-module $A \otimes E \subset C(X;E)$, f belongs to the uniform closure of $A \otimes E$ in $C(X;E)$. Since $A \otimes E \subset W$, the proof is complete.

COROLLARY 9.46. *Let X be a 0-dimensional compact Hausdorff space, and let E and W be as in Theorem 9.45. The following statements are equivalent.*

(1) W *is uniformly dense in* $C(X;E)$;

(2) $W(x,y) = \{(g(x),g(y));\ g \in W\}$ *is dense in* $X \times X$, *for every pair* $x, y \in X$;

(3) (a) *If* $x \neq y$, *there is* $g \in W$ *such that* $g(x) \neq g(y)$;

(b) *Given* $x \in X$, *there is* $g \in W$ *with* $g(x) \neq 0$.

(4) *Let* $A = \{u(g);\ u \in E', g \in W\}$. *Then* A *is separating over* X *and* $W(x) = \{g(x);\ g \in W\} = E$ *for every* $x \in X$.

PROOF: (1) $\Longrightarrow$ (2) $\Rightarrow$ (3) are immediate from Theorem 9.45.(3)$\Rightarrow$(4) follows from the hypothesis that E' is separating over E and from $A \otimes E \subset W$.

Finally, (4) $\Rightarrow$ (1) by Corollary 9.22 applied to the A-module $A \otimes E$, which is contained in W.

COROLLARY 9.47. (Weierstrass polynomial approximation) *Let* E_1 *and* E_2 *be two nonarchimedean normed spaces over* F *such that* E_i' *is separating over* E_i $(i = 1,2)$. *For every compact subset* $K \subset E_1$ *the set* $P_f(E_1 ; E_2)|K$ *is uniformly dense in* $C(K;E_2)$.

PROOF: Let $W = P_f(E_1 ; E_2)|K$. Since E_2' is separating over E_2, W is a polynomial algebra contained in $C(K;E_2)$. Now W contains the constants and it is separating over K, because E_1' is separating over E_1. It remains to apply the preceding Corollary.

As another application of the general results proved above, let us give a nonarchimedean analogue of Blatter's Stone-Weierstrass Theorems for finite-dimensional non-associative real algebras (see Theorems 1.22 and 1.24 of [4]).

Let E be a finite-dimensional non-associative (i.e. not necessarily associative) linear algebra over a *complete* non-archimedean non-trivially valued field F. Since every field provided with a topology induced by a non-trivial valuation is *strictly minimal* (see Nachbin [75]), there is a unique Hausdorff topology on E that makes it a topological vector space over F, and moreover, under this topology, every linear transformation $T : E \to E$ is continuous. (See Nachbin [75], Theorems 7 and 9.) We shall always consider E endowed with its unique Hausdorff topology that makes it a topological vector space over F. This topology, called *admissible* in [75], can be defined as follows. If $\{e_1, \ldots, e_n\}$ is a basis of E over F, then the nonarchimedean sup-norm

$$\| v \| = \max \{ |v_i| ; 1 \leq i \leq n\},$$

whenever $v = \sum_{i=1}^{n} v_i e_i$ is in E, defines the unique admissible topology of E .

If we define operations pointwise, $C(X;E)$ becomes a non - associative *algebra* over F too, as well a *bimodule* over E: if $v \in E$ and $f \in C(X;E)$ then the mappings $x \to vf(x)$ and $x \to f(x)v$ belong to $C(X;E)$. A vector subspace $W \subset C(X;E)$ is called a *submodule over* E if is a bimodule over E , with the above operations. An algebra E is called a *zero - algebra* if $uv = 0$ for all $u, v \in E$. The algebra E is called *simple* if it is not a zero - algebra and has no subspaces invariant relative to the right and left multiplications, except 0 and E. Let $\mathcal{M}(E)$ be the subalgebra of $\mathcal{L}(E)$ generated by the set of all right and left multiplications. $\mathcal{M}(E)$ is called the *multiplication algebra* of E . It follows that a non - zero - algebra is simple if, and only if, $\mathcal{M}(E)$ is a irreducible algebra of linear transformations. The *centroid* of E is the set of all linear transformations $T \in \mathcal{L}(E)$ which commute with all right and left multiplications. Clearly, all linear transformations of the form λI belong to the centroid of E , where $\lambda \in F$ and I is the identity map of E . We say that E is *central* if its centroid is just $\{\lambda I ; \lambda \in F\}$.

THEOREM 9.48. *Let* F *be a complete and non - trivially valued nonarchimedean field. Let* E *be a finite - dimensional central and simple non - associative algebra over* F . *Let* $W \subset C(X;E)$ *be a subalgebra which is a submodule over* E . *Then, for every* $f \in C(X;E)$, *conditions* (1) - (4) *of Theorem 9.45 are equivalent.*

PROOF: The proof consists in showing that, under the above hypothesis on E , any subalgebra $W \subset C(X;E)$ which is a submodule over E is a polynomial algebra.

By Theorem 4, Chapter X , Jacobson [31], we have $\mathcal{M}(E) = \mathcal{L}(E)$. Hence the submodule W is invariant under composition with any linear transformation $T \in \mathcal{L}(E)$. Let

$$A = \{u(f);\ u \in E',\ f \in W\} .$$

By Lemma 4.1, extended to the case of F , A is a vector subspace of $C(X;F)$ and $A \otimes E \subset W$. It remains to prove that A is closed under multiplication. Since E is not a zero-algebra, choose a pair u_o, v_o in E such that $u_o v_o \neq 0$. Let $u \in E'$ such that $u(u_o v_o) = 1$. Let $v(f)$ and $w(g)$ be in A. The mappings

$$x \to v(f(x))u_o \quad \text{and} \quad x \to w(g(x))v_o$$

belong to W, since $A \otimes E \subset W$. By hypothesis, W is a subalgebra of $C(X;E)$. Therefore,

$$x \to [v(f(x))u_o] \cdot [w(g(x))v_o] = v(f(x))\, w(g(x)u_o v_o$$

belongs to W. Call it h. Then $u(h) \in A$, and $u(h) = v(f)w(g)$, since $u(u_o v_o) = 1$. Thus W is polynomial algebra.

REMARK. Notice that a subalgebra $W \subset C(X;E)$ which contains the constants is a submodule over E. Conversely, if W is a unitary subalgebra which is a submodule over E, then W contains the constants.

§ 7. BISHOP'S THEOREM.

Let F be a nonarchimedean valued field. If F is *complete*, and K is a finite extension of F, then the rank one valuation $t \to |t| \in \mathbb{R}_+$ of F can be extended from F to K in a unique way as a rank one valuation. If F is not complete, then its valuation can be extended to a rank one valuation of K in finitely many non-equivalent ways.

DEFINITION 9.49. *Let F be a nonarchimedean valued field; let K be a finite algebraic extension of F, endowed with a rank one valuation extending that of F. Let $A \subset C(X;K)$ be a subalgebra. A subset $S \subset X$ is called A-antisymmetric (with respect to F) if, for every $a \in A$, $a|S$ being F-valued implies that $a|S$ is constant.*

DEFINITION 9.50. *Let $x, y \in X$. We write $x \equiv y$ if there is an A-antisymmetric set S which contains both x and y.*

The equivalence classes modulo the equivalence relation $x \equiv y$ are called *maximal* A-*antisymmetric sets (with respect to* F).

The following result is the nonarchimedean analogue of Machado's version of Bishop's Theorem [37]. In it, F is a nonarchimedean valued field; K is a finite algebraic extension of F, and K is valued by one extension to K of the valuation of F; X is a compact Hausdorff space and E is a nonarchimedean normed space over K.

THEOREM 9.51. *Let* $A \subset C(X;K)$ *be a subalgebra; let* $W \subset C(X;E)$ *be a vector subspace which is an* A-*module. For each* $f \in C(X;E)$, *there is a maximal* A-*antisymmetric set (with respect to* F) $S \subset X$ *such that*

$$d(f\,;W) = d(f|S\,;W|S).$$

COROLLARY 9.52. *Let* A *and* W *be as in Theorem 9.51, and let* $f \in C(X;E)$. *Then* f *belongs to the closure of* W *in* C(X;E) *if, and only if,* $f|S$ *belongs to the closure of* $W|S$ *in* C(S;E), *for each maximal* A-*antisymmetric set (with respect to* F) $S \subset X$.

PROOF OF THEOREM 9.51. Let $f \in C(X;E)$. Put $d = d(f,W)$. We can assume $d > 0$, the result being clear for $d = 0$, since $d(f|S\,;W|S) \leq d$ for any $S \subset X$. Let D be the set of all ordered pairs (P, S) such that

(i) P is a partition of X into non-empty pairwise disjoint and closed subsets of X;

(ii) $S \in P$ and $d = d(f|S\,;W|S)$.

The pair ({X}, X) belongs to D, so $D \neq \emptyset$. We partially order D by setting $(P,S) \leq (Q,T)$ if, and only if, the partition Q is finer than P, and $T \subset S$. The arguments in Machado's proof of Bishop's Theorem (see [37]) apply here, so that each chain in D has an upper bound. By Zorn's Lemma there is a maximal element $(Q,T) \in D$. We claim that T is A-antisymmetric (with respect to F). Indeed, let A_T be the set $\{a \in A\,;\ a|T$ is F-valued$\}$. By contradiction admit that

$B = A_T|T$ contains non-constant functions. Since $B \subset C(T;F)$, and $W|T$ is a B-module, by Theorem 9.21 we may find an equivalence class $V \subset T$ (modulo T/B) such that

$$d(f|T ; W|T) = d(f|V ; W|V).$$

Since $d = d(f|T ; W|T)$, and V is proper subset of T, the partition P of X consisting of the elements of Q distinct from T and by the equivalence classes of T modulo T/B is strictly finer then Q, and therefore $(Q ; T) < (P , V)$, which contradicts the maximality of (Q , T). The *maximal* A-antisymmetric set S, which contains T, is then such that $d = d(f|S ; W|S)$.

§ 8. TIETZE EXTENSION THEOREM

Let Y be a closed non-empty subset of a 0-dimensional compact Hausdorff space X, and let E be a nonarchimedean *Banach space* over a valued field $(F ; |\cdot|)$ (i.e., E is a nonarchimedean normed space over F which is *complete*). Let

$$T_Y : C(X ; E) \to C(Y ; E)$$

be the restriction map defined as follows:

$$T_Y(f) = f|Y$$

for all $f \in C(X;E)$. This map is obviously linear and continuous, since

$$\|T_Y(f)\| = \|f|Y\| \leq \|f\|$$

for all $f \in C(X;E)$.

Let $C(X;E)|Y$ be the image of $C(X;E)$ under T_Y in $C(Y;E)$.

THEOREM 9.53. *The vector subspace* $C(X;E)|Y$ *is uniformly dense in* $C(Y;E)$.

PROOF: Let us define $A = \{f \in C(Y;F);\ f = g|Y \text{ for some } g \in C(X;F)\}$.

Clearly, A is a unitary subalgebra of $C(Y;F)$. Since X is 0-dimensional, A separates the points of Y. Let $W = C(X;E)|Y$. Clearly, W is an A-module. Moreover, $W(x) = E$ for all $x \in X$, since W contains the constants. By Corollary 9.45, W is uniformly dense in $C(Y;E)$, as claimed.

DEFINITION 9.54. *A topological space* X *is called ultranormal if, given any two closed disjoint sets* A *and* B *in* X, *there exists two clopen disjoint sets* U *and* V *in* X *such that* $A \subset U$ *and* $B \subset V$.

LEMMA 9.55. *Let* X *be a topological space. Then* X *is ultranormal if, and only if, given a pair of closed disjoint sets* A *and* B *in* X, *and a nonarchimedean valued field* $(F, |\cdot|)$, *there is an element* $f \in C(X;F)$ *such that* $f(A) = \{0\}$, $f(B) = \{1\}$ *and* $|f(x)| \leq 1$ *for all* $x \in X$.

PROOF: Let X be a ultranormal topological space, let A and B be a pair of closed disjoint sets in X, and let $(F, |\cdot|)$ be a nonarchimedean valued field. Let U and V be two clopen disjoint sets such that $A \subset U$ and $B \subset V$. Since V is clopen, the F-valued characteristic function $\phi_V : X \to F$ is continuous. Clearly, $\phi_V(A) = \{0\}$, because $A \subset U$ and $U \cap V = \emptyset$. Since $B \subset V$, $\phi_V(B) = \{1\}$. Finally, $\phi_V(X) = \{0, 1\}$. Hence $|\phi_V(x)| \leq 1$ for all $x \in X$.

Conversely, assume X is such that, given a pair of closed disjoint sets A and B in X, and a nonarchimedean valued field $(F, |\cdot|)$ there is an element $f \in C(X;F)$ such that $f(A) = \{0\}$, $f(B) = \{1\}$ and $|f(x)| \leq 1$ for all $x \in X$. Let us define

$$U = \{t \in X;\ |f(t)| < 1\}$$

and

$$V = \{t \in X;\ |f(t)| \geq 1\}.$$

Then $U = f^{-1}(B_1(0))$. Since $B_1(0) = \{v \in f;\ |v| < 1\}$ is clopen, U is clopen. Clearly $A \subset U$. On the other hand, the set $S = F \setminus B_1(0)$ is clopen too, and $V = f^{-1}(S)$. Therefore V is clopen, $B \subset V$ and the intersection $U \cap V = \emptyset$. This ends the proof.

LEMMA 9.56. *Every compact Hausdorff* 0 - *dimensional space is ultranormal.*

PROOF: Let X be a 0 - dimensional compact Hausdorff space. Let A and B be two closed disjoint sets on X. Let $b \in B$. For each $a \in A$, there is a clopen set U_a in X and a clopen set V_a in X such that $a \in U_a$, $b \in V_a$, $U_a \cap V_a = \emptyset$. By compactness of A, there are finitely many $a_1, a_2, \ldots, a_n \in A$ such that A is contained in the union $U_b = U_{a_1} \cup \ldots \cup U_{a_n}$. Clearly, U_b is a clopen set in X, with $A \subset U_b$. Let $V_b = V_{a_1} \cap \ldots \cap V_{a_n}$. Then V_b is a clopen neighborhood of b. By compactness of B, there are finitely many $b_1, b_2, \ldots, b_m \in B$ such that B is contained in the union $V = V_{b_1} \cup \ldots \cup V_{b_m}$. Clearly, V is a clopen set in X with $B \subset V$. Let $U = U_{b_1} \cap \ldots \cap U_{b_m}$. Clearly, U is a clopen set in X with $A \subset U$. Finally, one easily sees that $U \cap V = \emptyset$.

THEOREM 9.57. *Let* X *be a compact Hausdorff* 0 - *dimensional space. For any closed non - empty subset* $Y \subset X$, *the continuous linear mapping* $T_Y : C(X;E) \to C(Y;E)$ *is a topological homomorphism, for each nonarchimedean normed space* E.

PROOF: Let us consider the neighborhood base of 0 in $C(X;E)$ consisting of all subsets of the form

$$N = \{g \in C(X;E);\ \|g(x)\| < \varepsilon,\ x \in X\},$$

for $\varepsilon > 0$. We have to prove that for each such N, the image $T_Y(N)$ is relatively open in $C(X;E)|Y = T_Y(C(X;E))$.

Let then, for each such N, define

$$W = \{h \in C(Y;E);\ \|h(x)\| < \varepsilon,\ x \in Y\}.$$

This is an open neighborhood of 0 in $C(Y;E)$. We claim that

$$T_Y(N) = W \cap T_Y(C(X;E)),$$

whence $T_Y(N)$ is relatively open in the image $T_Y(C(X;E)) = C(X;E)|Y$. The inclusion

(i) $T_Y(N) \subset W \cap T_Y(C(X;E))$ is obvious. Conversely, let $h \in W \cap T_Y(C(X;E))$. Let $g \in C(X;E)$ be such that $h = T_Y(g)$. Therefore $g(x) = h(x)$ for all $x \in Y$. Define

$$B = \{t \in X; \|g(t)\| \geq \varepsilon\}.$$

Then $B \subset X$ is closed and disjoint from Y. Indeed, $h \in W$ implies that $\|h(x)\| < \varepsilon$, for all $x \in Y$. Hence $\|g(x)\| < \varepsilon$ for all $x \in Y$. If $B = \emptyset$, then $g \in N$, and therefore $h \in T_Y(N)$. If $B \neq \emptyset$, by Lemma 9.55, there is $f \in C(X;F)$ such that $f(B) = \{0\}$, $f(Y) = \{1\}$ and $|f(t)| \leq 1$ for all $t \in X$. We can apply Lemma 9.55, because by Lemma 9.56, X is ultranormal. Let $k = fg$. Then $k \in C(X;E)$, $k(x) = f(x)g(x) = h(x)$, if $x \in Y$. Therefore $h = T_Y(k)$. We claim that $k \in N$. Let $t \in X$. If $t \in B$, then $f(t) = 0$, so $k(t) = 0$ too. Therefore $|k(t)| < \varepsilon$. If $t \notin B$, then we have

$$\|k(t)\| = \|f(t)g(t)\| = |f(t)| \cdot \|g(t)\|$$

$$\leq \|g(t)\| < \varepsilon .$$

Therefore $\|k(t)\| < \varepsilon$ for all $t \in X$, and $k \in N$. This shows that $k \in N$, i.e. $h \in T_Y(N)$. This ends the proof that

(ii) $W \cap T_Y(C(X;E)) \subset T_Y(N)$.

From (i) and (ii), it follows the desired equality, and $T_Y(N)$ is relatively open, Q E D.

THEOREM 9.58. *Let* X *be a compact Hausdorff* 0 - *dimensional space, and let* $Y \subset X$ *be a non - empty closed subset. Then, for each nonarchimedean Banach space* E, *over valued field* $(F; |\cdot|)$, *we have:*

$$C(X;E)|Y = C(Y;E).$$

PROOF: By Theorem 9.53, all we have to prove is that $C(X;E)|Y$ is closed in $C(Y;E)$. Let K be the kernel of T_Y in $C(X;E)$, i.e. $K = \{f \in C(X;E);\ T_Y(f) = 0\}$. Since T_Y is a continuous linear mapping, the kernel K is a closed subspace of the nonarchimedean Banach space $C(X;E)$. Hence the quotient space $C(X;E)/K$ is a Banach space too, and therefore complete.

By Theorem 9.57, the mapping T_Y is a topological homomorphism. Hence $C(X;E)/K$ and $T_Y(C(X;E)) = C(X;E)|Y$ are topologically linearly isomorphic. Thus $C(X;E)|Y$ is complete, and therefore closed in $C(Y;E)$.

REMARK 9.59. When $E = F$ and F is the field of p-*adic numbers*, i.e. the completion of field Q with the p-adic valuation defined in Example 9.3, with the extension of $|\cdot|_p$ from Q to F, then Theorem 9.58 is due to J. Dieudonné (see Théorème 1, [70]).

REMARK 9.60. When $E = F$ and F is a locally compact (hence complete) nonarchimedean nontrivially valued field, then Theorem 9.58 is valid without the hypothesis that X be compact. It is enough to assume that X is ultranormal, and then the conclusion is that

$$C_b(X;F)|Y = C_b(Y;F)$$

for any closed subset $Y \subset X$. This version of the Tietze extension Theorem is due to R. L. Ellis, A nonarchimedean analog of the Tietze-Urysohn extension Theorem, *Indagationes Math.*, 70, p. 332-333.

§ 9. THE COMPACT - OPEN TOPOLOGY.

DEFINITION 9.61. *Let* E *be a vector space over a valued field* $(F,|\cdot|)$. *A mapping* $p: E \to \mathbb{R}$ *is called a seminorm on* E *if*

(1) $p(x) \geq 0$ *for all* $x \in E$;

(2) $p(\lambda x) = |\lambda| \cdot p(x)$, *for all* $\lambda \in F$, $x \in E$,

(3) $p(x + y) \leq p(x) + p(y)$, *for all* $x, y \in E$.

If, moreover, the property.

(4) $p(x + y) \leq \max(p(x), p(y))$, *for all* $x, y \in E$ *is true, we say that* p *is a nonarchimedean seminorm on* E.

Let E be a vector space over a nonarchimedean valued field $(F, |\cdot|)$. Let Γ be a family of nonarchimedean seminorms on E. We define a topology τ on E by setting as a basis of neighborhoods of 0 the sets of the form

$$\{x \in E \,;\, p_i(x) \leq \varepsilon\,,\ i = 1,2,\ldots,n\}$$

where $p_i \in \Gamma$, $i = 1,2,\ldots,n$, and $\varepsilon > 0$. We say that the topology τ is determined by the family Γ. Then (E,τ) is a *topological vector space* over F, i.e. the following is true

(i) the map $(x, y) \mapsto x + y$ of $E \times E$ into E is continuous;

(ii) the map $(\lambda, x) \mapsto \lambda x$ of $F \times E$ into E is continuous.

DEFINITION 9.62. *Let* E *be a vector space over a nonarchimedean valued field* $(F, |\cdot|)$. *A subset* $X \subset E$ *is said to be* F*-convex if the following is true:*

$$\alpha x + \beta y + \gamma z \in X \quad \textit{for all} \quad x, y, z \in X$$

$$\textit{and} \quad \alpha,\beta,\gamma \in F \quad \textit{such that} \quad |\alpha| \leq 1, |\beta| \leq 1,$$

$$|\gamma| \leq 1, \textit{ and } \quad \alpha + \beta + \gamma = 1$$

REMARK. For every nonarchimedean seminorm p on a vector space E over $(F, |\cdot|)$, the following sets are convex $\{x \in E; p(x - x_0) < \varepsilon\}$ and $\{x \in E; p(x - x_0) \leq \varepsilon\}$, for all $x_0 \in E$, and $\varepsilon > 0$.

DEFINITION 9.63. *Let* E *be a vector space over a nonarchimedean valued field* $(F, |\cdot|)$. *A topology* τ *that makes* (E, τ) *a topological vector space over* F *is said to be locally* F*-convex if there exists a fundamental system of neighborhoods of* 0 *consisting of* F*-convex sets.*

It follows that any topology determined by a family Γ of seminorms on E is locally F-convex. Conversely, one can show that the converse is also true. (See Monna [73]).

DEFINITION 9.64. *Let* X *be a Hausdorff space. Let* $(F,|\cdot|)$ *be a nonarchimedean valued field. For every compact subset* $K \subset X$, *let*

$$P_K(f) = \sup\{|f(x)| \; ; \; x \in K\}$$

for $f \in C(X;F)$.

One easily verifies that P_K is a nonarchimedean seminorm on $C(X;F)$. The *compact-open topology* on $C(X;F)$ is the locally F-convex topology determined by the family of seminorms

$$\Gamma = \{P_K \; ; \; K \subset X \text{ compact}\}.$$

More generally, if E is a nonarchimedean locally F-convex space, whose topology is determined by a family Γ of nonarchimedean seminorms on E, one defines a corresponding family of seminorms on $C(X;E)$ by setting

$$f \mapsto \sup\{p(f(x)); \; x \in K\}$$

for $f \in C(X;E)$, $p \in \Gamma$ and $K \subset X$ a compact subset. This is the compact-open topology on $C(X;E)$. In particular, when E is a nonarchimedean normed algebra, with norm $t \mapsto \|t\|$, the seminorms

$$P_K(f) = \sup\{\|f(x)\| \; ; \; x \in K\}$$

on $C(X;E)$ have the property

(i) $\quad P_K(fg) \leq P_K(f) \cdot P_K(g)$

for all $f, g \in C(X;E)$. If E is unitary, with unit e, $\|e\| = 1$, then the constant function $x \mapsto e$, still denoted by e, is the unit of $C(X;E)$, and for every compact subset $K \subset X$, one has

(ii) $P_K(e) = 1$

In view of properties (i) and (ii), one says that the seminorms P_K are *algebra seminorms*. As a Corollary multiplication in $C(X;E)$ is continuous.

Therefore $C(X;E)$ with the compact-open topology is termed a *nonarchimedean topological algebra*. It is easy to see that the closure of an algebra, or of a right (resp. left) ideal in $C(X;E)$ is also a subalgebra or a right (resp. left) ideal in $C(X;E)$. The problem arises of characterizing the compact - open closure of a subalgebra or of a right (resp. left) ideal in $C(X;E)$, and in particular in $C(X;F)$, since by property (3) of Definition 9.1, any nonarchimedean valued field $(F, |\cdot|)$ is a unitary nonarchimedean normed algebra over itself.

THEOREM 9.65. *Let* X *be a Hausdorff space. Let* E *be a unitary nonarchimedean normed algebra over a valued field* $(F, |\cdot|)$. *Let* $A \subset C(X;E)$ *be a separating (in the sense of Definition 9.31) unitary subalgebra of* $C(X;E)$, *and let* $W \subset C(X;E)$ *be a vector subspace which is an* A*-module. Then* W *is local.*

Before proving Theorem 9.65 let us define what we mean by saying that W is local.

DEFINITION 9.66. *Let* X *be a Hausdorff space, and let* E *be a nonarchimedean normed space over a valued field* $(F, |\cdot|)$. *Let* $W \subset C(X;E)$. *We say that* W *is local if any* $f \in C(X;E)$ *which is in* W *locally at all points of* X *is then in the compact - open closure of* W.

Notice that, since $\{x\} \subset X$ is compact, for any $x \in X$, all functions f in $\overline{W}$, the compact-open closure of W, are in W locally at all points of X, i.e., $\tilde{\Delta}(W) \supset \overline{W}$. Therefore W is local if and only if $\tilde{\Delta}(W) = \overline{W}$.

PROOF OF THEOREM 9.65. Let $f \in C(X;E)$ be in $\tilde{\Delta}(W)$. Let $K \subset X$ and $\varepsilon > 0$ be given. Then $f|K$ is in $C(K;E)$; $A|K \subset C(K;E)$ is a separating unitary subalgebra of $C(K;E)$; and $W|K$ is an $(A|K)$ - module. Since $f|K \in \Delta(W|K)$, and by Theorem 9.35, $W|K$ is local, then $f|K$ belongs to the uniform closure of $W|K$ in $C(K;E)$. Therefore a $g \in W$ can be found such that

$$\|g(x) - f(x)\| < \varepsilon$$

for all $x \in K$, i.e. $P_K(f - g) < \varepsilon$.

COROLLARY 9.66. *Let* X, E, A *and* W *be as in Theorem 9.65. Then* $f \in C(X;E)$ *is in the compact - open closure of* W *in* $C(X;E)$ *if, and only if,* $f(x) \in \overline{W(x)}$ *in* E, *for each* $x \in X$.

PROOF: Since each $\{x\} \subset X$ is compact, the condition is obviously necessary. Conversely, if the condition is verified, then given $x \in X$ and $\varepsilon > 0$, there is $g \in W$ such that

$$\|f(x) - g(x)\| < \varepsilon .$$

By continuity this is still true in a neighborhood U of x in X. Thus $f \in \tilde{\Delta}(W)$. By Theorem 9.65, $f \in \overline{W}$, as desired.

COROLLARY 9.67. *Let* X, E, A *and* W *be as in Theorem 9.65. Assume that* W *contains the constants. Then* W *is dense in the compact - open topology of* $C(X;E)$.

PROOF: Apply Corollary 9.66, noticing that $W(x) = E$, for all $x \in X$.

COROLLARY 9.68. *Let* X *and* E *be as in Theorem 9.65. Assume* X *is 0 - dimensional and let* $I \subset C(X;E)$ *be a closed right (resp. left) ideal, and for each* $x \in X$, *let* I_x *be the closure in* E *of the set*

$$I(x) = \{f(x);\ f \in I\} .$$

then I_x *is a closed right (resp. left) ideal in* E, *and*

$$I = \{f \in C(X;E);\ f(x) \in I_x \text{ for all } x \in X\} .$$

PROOF: The fact that

$$I = \{f \in C(X;E);\ f(x) \in I_x \ \text{ for all } \ x \in X\}$$

follows from Corollary 9.66 and the hypothesis that I is closed, if we can show that $C(X;E)$ is a unitary separating subalgebra in the sense of Definition 9.31. An analysis of the proof of Theorem 9.65 shows that in fact all we need to prove is that $C(X;E)|K$ is separating in $C(K;E)$, for all compact subsets $K \subset X$. Now X is 0-dimensional, therefore $C_b(X;F)$ separates points in the sense that given $x \neq y$ in X there is some $f \in C_b(X;F)$ such that $f(x) = 1$ and $f(y) = 0$. Now $f(K)$ is compact in F. By Kaplansky's Lemma, there is a polynomial $p : F \to F$ such that $p(1) = 1$, $p(0) = 0$ and $|p(t)| \leq 1$ for all $t \in f(K)$. Let us define $h = p \circ f$. Then $g = h \otimes e$ belongs to $C(X;E)$, $g(x) = e$, $g(y) = 0$ and $\|g(y)\| \leq 1$ for all $y \in K$. This shows that $C(X;E)|K$ is separating in the sense of Definition 9.31.

The proof that $I(x)$ is a right (resp. left) ideal in E, for each $x \in X$, is easy. Then I_x, being its closure, is a closed right (resp. left) ideal in E follows from the fact that E is a topological algebra.

COROLLARY 9.69. *Let* X *and* E *be as in Corollary 9.68. Assume that* E *is simple. Then any closed two-sided ideal in* $C(X;E)$ *consists of all functions vanishing on a closed subset of* X. *Moreover, any maximal two-sided closed ideal in* $C(X;E)$ *is of the form* $\{f \in C(X;E);\ f(x) = 0\}$ *for some point* $x \in X$.

PROOF: The proof is similar to the case of X compact and the uniform topology, so we omit the details.

§10. THE NONARCHIMEDEAN STRICT TOPOLOGY.

In this section X is a *locally compact* Hausdorff space, and E is a nonarchimedean normed space over a locally compact valued field $(F, |\cdot|)$. On the vector space $C_b(X;E)$ of all bounded continuous E-valued functions let us define a locally F-convex topology β, called the *strict topology*, by setting

$$P_\phi(f) = \sup\{ \|\phi(x)\ f(x)\| \ ; \ x \in X\}$$

for all $f \in C_b(X;E)$, where $\phi \in C_o(X;F)$. Here $C_o(X;F)$ denotes the vector subspace of $C(X;F)$ consisting of all those $\phi \in C(X;F)$ such that, given $\varepsilon > 0$ there is a *compact* subset $K \subset X$ such that $|\phi(x)| < \varepsilon$ for all $x \in X$ outside of K. It follows that $\{t \in X;\ |\phi(t)| \geq \varepsilon\}$ is compact and open for every $\varepsilon > 0$.

THEOREM 9.70. *Let* $A \subset C_b(X;F)$ *be a separating unitary subalgebra, and let* $W \subset C_b(X;E)$ *be a vector subspace which is an* A - *module. Then* W *is* β - *local.*

Before proving Theorem 9.70 let us define what we mean by saying that W is β - local.

DEFINITION 9.71. *If* $W \subset C_b(X;E)$, *we say that* W *is* β - *local if any* $f \in C_b(X;E)$ *which is in* W *locally at all points of* X *is then in the strict closure of* W *in* $C_b(X;E)$.

Since, for each point $x \in X$, there is an open and compact neighborhood K of x in X, the F - characteristic function ϕ_K of K is such that $\phi_K(x) = 1$ and $\phi_K \in C_o(X;F)$. Hence, all functions f in $\overline{W}$, the β - closure of W is $C_b(X;E)$ are in W locally at all points of X, i.e. $\tilde{\Delta}_b(W) \supset \overline{W}$. Therefore W is β - local if, and only if, $\tilde{\Delta}_b(W) = \overline{W}$, the β - closure of W in $C_b(X;E)$. (Here $\tilde{\Delta}_b(W) = \tilde{\Delta}(W) \cap C_b(X;E)$).

LEMMA 9.72. *Let* A *be as in Theorem 9.70. For every* $x \in X$, *let there be given a compact subset* $K_x \subset X$, *not containing* x. *Then there exist a finite set* $x_1, x_2, \ldots, x_n \in X$ *and functions* $\phi_1, \phi_2, \ldots, \phi_n$ *in the uniform closure of* A *such that* $\phi_i(t) = 0$ *for all* $t \in K_{x_i}$ *for* $i = 1,2,\ldots,n$; $\phi_1 + \ldots + \phi_n = 1$ *in* X *and* $|\phi_i(t)| \leq 1$ *for all* $t \in X$, $i = 1,2,\ldots,n$.

PROOF: Introduce the nonarchimedean Stone - Čech compactification $\beta_F X \supset X$. This is done as follows. Since F is locally compact, the sets $V_r = \{a \in F;\ |a| \leq r\}$ are compact, for every

$r \in \mathbb{R}$, $r > 0$. Now each $f \in C_b(X;F) = H$ is such that $f(X) \subset V_{r_f}$ for some $r_f > 0$. Consider the map $e : X \to \prod_{f \in H} V_{r_f}$ define by

$$x \mapsto (f(t))_{f \in H} \;;$$

since the space X is a 0 - dimensional Hausdorff space, this mapping is a topological embedding, and $\beta_F X$ is the closure of $e(X)$ in $\prod_{f \in H} V_{r_f}$.

As in the classical case, each $f \in C_b(X;F)$ has a unique continuous F - valued extension βf to $\beta_F X$. The mapping $\beta\, C_b(X;F) \to C(\beta_F X;F)$ defined by $f \mapsto \beta f$, is then a Banach algebra isomorphism between $C_b(X;F)$ and $C(\beta_F X;F)$. Let $B = \beta A$. Consider the quotient space Y of βX modulo the equivalence relation B, and let $\pi : \beta_F X \to Y$ be the quotient map. Then Y is a compact 0 - dimensional Hausdorff space. If $x \in X$, then $\pi(x) \cap X$ is an equivalence class in X modulo X/A. Therefore $\pi(x) \cap X = \{x\}$, and $\pi(x) \cap X$ is disjoint from K_x. Thus $\pi(x) \notin \pi(K_x)$. Hence

$$\bigcap_{x \in X} \pi(K_x) = \emptyset.$$

By the finite intersection property, there is a finite set $\{x_1, x_2, \ldots, x_n\} \subset X$ such that

$$\pi(K_{x_1}) \cap \ldots \cap \pi(K_{x_n}) = \emptyset .$$

By Lemma 9.18, there exist functions $h_i \in C(Y;F)$, $i = 1,2,\ldots,n$, such that

(a) $h_i(y) = 0$ for all $y \in \pi(K_{x_i})$ $(i = 1,\ldots,n)$

(b) $\|h_i\| \leq 1$, for all $i = 1,2,\ldots,n$.

(c) $h_1 + h_2 + \ldots + h_n = 1$ on Y.

Put $\psi_i = h_i \circ \pi$, $i = 1,2,\ldots,n$. By Kaplansky's Theorem, ψ_i belongs to the uniform closure of B in $C(\beta_F X;F)$. It is clear that $\phi_i = \psi_i | X$, $i = 1,2,\ldots,n$ belong to the uniform closure of A in $C_b(X;F)$ and have all the desired properties.

PROOF OF THEOREM 9.70. Let $f \in \tilde{\Delta}_b(W) = \tilde{\Delta}(W) \cap C_b(X;E)$. Let $\phi \in C_o(X;F)$ and $\varepsilon > 0$ be given. We may assume $\|\phi\| > 0$.

For each $x \in X$, there is $g_x \in W$ and neighborhood U_x of x in X such that

$$\|f(t) - g_x(t)\| < \varepsilon / \|\phi\|$$

for all $t \in U_x$. In particular

$$\|\phi(x)(f(x) - g_x(x))\| < \varepsilon .$$

Let $K_x = \{t \in X; \|\phi(t)(f(t) - g_x(t))\| \geq \varepsilon\}$. Then K_x is compact and $x \notin K_x$. By Lemma 9.72 there exist $x_1, x_2, \ldots, x_n \in X$ and $\phi_1, \phi_2, \ldots, \phi_n \in C_b(X;F)$ belonging to the uniform closure of A in $C_b(X;F)$ such that

(a) $\phi_i(t) = 0$ for all $t \in K_{x_i}$ $(i = 1,2,\ldots,n)$;

(b) $|\phi_i(t)| \leq 1$ for all $t \in X$ $(i = 1,2,\ldots,n)$;

(c) $\phi_i + \phi_2 + \ldots + \phi_n = 1$ on X.

For each $i = 1,2,\ldots,n$ choose $h_i \in A$ such that

(d) $|\phi_i(t) - h_i(t)| < \varepsilon / P_\phi(g_{x_i})$ for all $t \in X$.

Define $g, h \in C_b(X;E)$ by

$$g = \sum_{i=1}^{n} \phi_i g_{x_i}, \quad h = \sum_{i=1}^{n} h_i g_{x_i}$$

We claim that

(e) $P_\phi(f - g) \leq \varepsilon$, $P_\phi(g - h) \leq \varepsilon$.

Indeed, take $t \in X$. Then

$$\|\phi(t)(f(t) - g(t))\| =$$

$$= \left\| \sum_{i=1}^{n} \phi(t)\ \phi_i(t)(f(t) - g_{x_i}(t)) \right\|$$

$$\leq \max_{1 \leq i \leq n} \{|\phi_i(t)| \cdot \|\phi(t)(f(t) - g_{x_i}(t))\|\}.$$

Now for those $i \in \{1,2,\ldots,n\}$ such that $t \in K_{x_i}$ we have $|\phi_i(t)| = 0$, and for those $i \in \{1,2,\ldots,n\}$ such that $t \notin K_{x_i}$, then $|\phi_i(t)| \leq 1$ while $\|\phi(t)(f(t) - g_{x_i}(t))\| < \varepsilon$. This shows that in any case

$$\|\phi(t)(f(t) - g_{x_i}(t))\| < \varepsilon ,$$

and therefore $P_\phi(f - g_{x_i}) \leq \varepsilon$.

On the other hand,

$$\|\phi(t)(g(t) - h(t))\| =$$

$$= \left\| \sum_{i=1}^{n} \phi(t)(\phi_i(t) - h_i(t))g_{x_i}(t) \right\|$$

$$\leq \max_{1 \leq i \leq n} \{|\phi_i(t) - h_i(t)| \cdot \|\phi(t)g_{x_i}(t)\|\}$$

$$\leq \max_{1 \leq i \leq n} \{|\phi_i(t) - h_i(t)| \cdot P_\phi(g_{x_i})\} < \varepsilon$$

Hence $P_\phi(g - h) \leq \varepsilon$.

From (e) it follows that $P_\phi(f - h) \leq \varepsilon$.

It remains to notice that $h \in W$, to conclude that f belongs the β - closure of W in $C_b(X;E)$.

COROLLARY 9.73. *Let* A *and* W *be as in Theorem 9.70. Then* $f \in C_b(X;E)$ *is in the* β *- closure of* W *in* $C_b(X;E)$ *if, and only if,* $f(x) \in \overline{W(x)}$ *in* E *for each* $x \in X$. *Moreover, if* W *contains the constants, then* W *is* β *- dense.*

PROOF: Since A is separating, X is 0 - dimensional. Let $x \in X$. Choose then a compact and open neighborhood U of x in X . The F - characteristic function of U , ϕ_U, belongs then to $C_0(X;F)$ and $\phi_U(x) = 1 > 0$. Hence the condition is necessary.

Conversely, let $f \in C_b(X;E)$ be such that $f(x) \in \overline{W(x)}$ in E for each $x \in X$. Thus given $x \in X$ and $\varepsilon > 0$, there is

$g \in W$ such that $\| f(x) - g(x) \| < \varepsilon$. By continuity this is still true in a neighborhood U of x in X. Thus $f \in \Delta_b(W)$. By Theorem 9.70, $f \in \overline{W}$, the β-closure of W in $C_b(X;E)$.

If W contains the constants, then $W(x) = E$ for all $x \in X$, and by the above argument, W is β-dense.

COROLLARY 9.74. *Let* X *be a* 0-*dimensional locally compact Hausdorff space. Let* $M \subset C_b(X;E)$ *be a* β-*closed* $C_b(X;F)$-*module and for each* $x \in X$, *let* M_x *be the closure in* E *of the set*

$$M(x) = \{f(x) ; f \in M\},$$

then M_x *is a closed vector subspace of* E *and*

$$M = \{f \in C_b(X;E) ; f(x) \in M_x \text{ for all } x \in X\}.$$

PROOF: Since X is 0-dimensional, the unitary algebra $C_b(X;F)$ is separating, and by the preceding corollary,

$$\overline{M} = \{f \in C_b(X;E) ; f(x) \in M_x \text{ for all } x \in X\}.$$

Since M is β-closed, $\overline{M} = M$. The fact that each M_x is a closed vector subspace of E is easy to establish.

COROLLARY 9.75. *Let* X *be as in Corollary 9.74 and let* E *be a unitary nonarchimedean normed algebra. Assume that* E *is simple. Then any* β-*closed two-sided ideal in* $C_b(X;E)$ *consists of all functions vanishing on a closed subset of* X. *Moreover, any maximal two-sided* β-*closed ideal in* $C_b(X;E)$ *is of the form* $\{f \in C_b(X;E) ; f(x) = 0\}$ *for some point* x *in* X.

PROOF: If $N \subset X$ is a closed subset, clearly

$$I(N) = \{f \in C_b(X;E);\ f(x) = 0 \text{ for all } x \in N\}$$

is a β-closed two-sided ideal. (Recall that given any $x \in X$, there is $\phi \in C_o(X;F)$ with $\phi(x) > 0$.)

If $I \subset C_b(X;E)$ is a β-closed two-sided ideal, define the closed set

$$N = \{t \in X;\ f(t) = 0 \quad \text{for all} \quad f \in I\}.$$

One easily sees that $I \subset I(N)$. To apply the preceding corollary we must show that I is a $C_b(X;F)$-module. Let $f \in I$ and $g \in C_b(X;F)$. Then $x \mapsto g(x)e$ belongs to $C_b(X;E)$. Call it h. Clearly $gf = hf \in I$.

Assume now that $f \in I(N)$, while $f \notin I$. By Corollary 9.74 there exists $x \in X$ such that $f(x) \notin I_x$. Hence $f(x) \neq 0$. Thus $x \notin N$, because $f \in I(N)$. On the other hand, since E is simple, either $I_x = \{0\}$ or $I_x = E$. Now $f(x) \notin I_x$ implies $I_x = \{0\}$. Thus $I(x) = 0$, i.e. $x \in N$, a contradiction. This shows $I(N) \subset I$.

REMARK 9.76. For further results on $C_b(X;E)$ with the strict topology β, and for more general nonarchimedean Nachbin spaces see the Doctoral Dissertation of José P. Carneiro, Universidade Federal do Rio de Janeiro, 1977. In fact, Theorem 9.70 is a Corollary of his result on localizability in the bounded case of the nonarchimedean Bernstein-Nachbin problem, dealing with not necessarily separating subalgebras $A \subset C_b(X;F)$.

8480-78-16
5-47

REFERENCES FOR CHAPTER 9.

MACHADO and PROLLA [39]

CHERNOFF, RASALA and WATERHOUSE [69]

DIEUDONNÉ [70]

INGLETON [71]

KAPLANSKY [72]

MONNA [73]

MURPHY [74]

NACHBIN [75]

NARICI, BECKENSTEIN and BACHMAN [76]

PROLLA [77]

B I B L I O G R A P H Y

[1] ARENS,R.; Approximation in, and representation of,certain Banach algebras, Amer.J.Math.71(1949), 763-790.

[2] ARENS,R.F., and KELLEY,J.L.; Characterizations of the space of continuous functions over a compact Hansdorff space, Trans. Amer. Math. Soc. 62 (1947), 499-508.

[3] ARON,R.M., and SCHOTTENLOHER,M., Compact holomorphic mappings on Banach spaces and the approximation property, J. Functional Analysis 21 (1976), 7-30.

[4] BLATTER,J., GROTHENDIECK spaces in approximation theory, Memoirs Amer. Math. Soc. 120 (1972).

[5] BIERSTEDT,K.D., Function algebras and a theorem of Mergelyan for vector-valued functions,in Papers from the Summer Gathering on Function Algebras at Aarhus, July 1969, Various Publications Series nº 9, Matematisk Institut, Aarhus Universitet, 1969.

[6] BIERSTEDT,K.D., The approximation property for weighted function spaces, Bonner Math. Schriften 81 (1975), 3-25.

[7] BIERSTEDT,K.D., and MEISE,R., Bemerkungen über die Approximationseigenschaft lokallconvexer Funktionenraüme, Math. Ann. 209 (1974), 99-107.

[8] BISHOP , E., A generalization of the Stone-Weierstrass theorem, Pacific J.Math. 11 (1961), 777-783.

[9] BRIEM,E.,LAURSEN,K.B., and PEDERSEN,N.W.,Mergelyan's theorem for vector-valued functions with an application to slice algebras, Studia Math.35 (1970), 221-226.

[10] BROSOWSKI,B., and DEUTSCH, F., On some geometric properties of suns, J. Approximation theory 10 (1974), 245-267.

[11] BUCK,R.C., Bounded continuous functions on a locally compact space, Michigan Math.J. 5 (1958), 95-104.

[12] BUCK,R.C., Approximation properties of vector-valued functions, Pacific J. Math. 53 (1974), 85-94.

[13] CHALICE,D.R., On a theorem of Rudin, Proc. Amer.Math.Soc. 35 (1972), 296-297.

[14] COLLINS,H.S., and DORROH,J.R., Remarks on certain function spaces Math.Ann. 176 (1968), 157-168.

[15] CUNNINGHAM,F.,JR., and ROY,N.M., Extreme functionals on an upper semicontimuous function space Proc. Amer.Math. Soc. 42 (1974), 461-465.

[16] DE LA FUENTE ANTÚNEZ, A.,Algunos resultados sobre aproximación de funciones vectoriales tipo teorema Weierstrass-Stone, Ph. D. Dissertation, Universidad de Madrid, 1973.

[17] DE LAMADRID,J.G., On finite dimensional approximations of mappings in Banach spaces, Proc.Amer.Math. Soc. 13 (1962), 163-168.

[18] DIEUDONNÉ,J., Sur les fonctions continues numériques définies daus un produit de deux espaces compacts, C.R Acad.Sci. Paris 205 (1937) 593-595.

[19] DUGUNDJI,J., An extension of Tietze's theorem, Pacific J. Math. 1 (1951), 353-367.

[20] DUNFORD,N., and SCHWARTZ,J.T., Linear Operator , vol I: General Theory, Pure and Applied Math. vol. 7, Interscience, New York, 1958.

[21] EIFLER,L., The slice product of function algebras, Proc. Amer.Math.Soc. 23 (1969), 559-564.

[22] ENFLO,P., A counterexample to the approximation problem in Banach spaces, Acta Math. 130 (1973),309-317.

[23] FIGIEL,T., Factorization of compact operators and applications to the approximation problem, Studia Math. 45 (1973), 191-210.

[24] FONTENOT,R.A., Strict topologies for vector-valued functions, Canadian J.Math. 26 (1974), 841-853.

[25] FREMLIN,D.H., GARLING,D.J.H., and HAYDON,R.G., Bounded Measures on topological spaces, Proc.London Math. Soc.(3) 25 (1972), 113-136.

[26] GLICKSBERG,I., Measures orthogonal to algebras and sets of antisymmetry Trans.Amer. Math. Soc. 105 (1962), 415-435.

[27] GLICKSBERG,I., Bishop's generalized Stone-Weierstrass theorem for the strict topology, Proc. Amer. Math. Soc. 14 (1963), 329-333.

[28] GROTHENDIECK,A., Espaces vectoriels topologiques, Universidade de São Paulo, São Paulo, Brazil, 1954.

[29] GROTHENDIECK,A., Sur certains espaces de fonctions holomorphes, I.,J. seine augenv. Math. 192 (1951), 35-64.

[30] HAYDON,R.G., On the Stone-Weierstrass theorem for the strict and superstrict topologies, Bonner Math. Schriften 81 (1975), 82 - 84.

[31] JACOBSON,N., Lie Algebras, Interscience New York, 1962.

[32] JEWETT,R.I., A variation on the Stone-Weierstrass theorem, Proc. Amer. Math. Soc. 14 (1963), 690-693.

[33] JOHNSON,W.B., Factoring compact operators, Israel J.Math. 9 (1971), 337-345.

[34] KLEINSTÜCK,G., Duals of weighted spaces of continuous functions, Bonner Math. Schriften 81 (1975), 98-114.

[35] KLEINSTÜCK,G., Der beschränkte **Fall des** gewichteten, Aproximationsproblems fur vektorwertige funktionen, Manuscripta Math. 17 (1975), 123-149.

[36] LANG,S., Algebra, Addison-Wesley, Reading, Man., 1967.

[37] MACHADO,S., On Bishop's generalization of the Weierstrass-Stone theorem, Indagationes Math., to appear.

[38] MACHADO,S., The Weierstrass-Stone theorem: a new proof, to appear.

[39] MACHADO,S., and PROLLA,J.B., An introduction to Nachbin spaces, Rendiconti del Circolo Matem. Palermo, Serie II, 21 (1972), 119-139.

[40] MACHADO,S., and PROLLA,J.B., The general complex case of the Bernstein-Nachbin approximation problem, Annales de l'Institut Fourier (Grenoble), to appear.

[41] MACHADO,S., and PROLLA,J.B., Concerning the bounded case of the Bernstein-Nachbin approximation problem, J. Math. Soc. Japan, to appear.

[42] NACHBIN,L., Weighted approximation for algebras and modules of continuous functions: real and self-adjoint complex cases, Annals of Math. 81 (1965) 289-302.

[43] NACHBIN,L., Elements of Approximation Theory, D. Van Nostrand Co., Inc., 1967. Reprinted ley R. Krieger Co., Inc., 1976.

[44] NACHBIN,L., Topology on spaces of holomorphic mappings, Erg. d. Math. 47, Springer-Verlag, Berlin, 1969.

[45] NACHBIN,L., On the priority of algebras of continuous functions in weighted approximation, Symposia Mathematica, to appear.

[46] NACHBIN,L., MACHADO,S., and PROLLA,J.B., Weighted Approximation, vector fibrations and algebras of operators, J. Math. pures et appl. 50(1971), 299-323.

[47] PEŁCZYŃSKI,A., A generalization of Stone's theorem on approximation, Bull. Acad. Polonaise Sci., Cl. III, 5 (1957), 105-107.

[48] PRENTER,P.M., A Weierstrass theorem for real separable Hilbert space, J. Approximation theory 3 (1970), 341-351.

[49] PRENTER,P.M., On polynomial operators and equations, in Nonlinear Functional Analysis and Applications (ed. L.B. Rall), Academic Press 1971, 361-398.

[50] PROLLA,J.B., The weighted Dieudonné theorem for density in tensor products, Indag.Math. 33 (1971), 170-175.

[51] PROLLA,J.B., Bishop's generalized Stone-Weierstrass theorem for weighted spaces, Math. Ann 191 (1971), 283-289.

[52] PROLLA,J.B., and MACHADO,S., Weighted Grothendieck subspaces, Trans. Amer. Math. Soc. 186 (1973), 247-258.

[53] RESTREPO,G., An infinite dimensional version of a theorem of Bernstein, Proc.Amer. Math. Soc. 23 (1969), 193-198.

[54] RUDIN, W., Subalgebras of spaces of continuous functions, Proc.Amer. Math. Soc. 7 (1956), 825-830.

[55] RUDIN, W., Real and complex analysis, Mc Graw-Hill Co., New York, 1966.

[56] RUDIN,W., Functional Analysis, Mc Graw-Hill Co., New York 1973.

[57] SCHAEFER,H.H., Topological Vector Spaces, Springer-Verlag Berlin, 1971.

[58] SCHAFER,R.D., An introduction to nonassociative algebras, Academic Press, New York and London, 1966.

[59] SCHWARTZ,L., Théorie des distributions a valeurs vectorielles I, Annales de l'Institute Fourier (1957) 1-141.

[60] SINGER,I., Best Approximation in Normed Linear Spaces by Elements of Linear Subspaces, Springer-Verlag, Berlin, 1970.

[61] STONE, M. H., On the compactification of topological spaces Ann. Soc. Polonaise Math. 21 (1948), 153 - 160.

[62] STONE, M. H., The generalized Weierstrass approximation theorem, in Studies in Modern Analysis (ed. R.C. Buck), MAA Studies in Math. 1 (1962), 30 - 87.

[63] STRÖBELE, W. J., On the representation of the extremal functionals on $C_0(T;X)$, J. approximation theory 10 (1974), 64 - 68.

[64] SUMMERS, W. H., Weighted approximation for modules of continuous function II, in Analyse Fonctionelle et Applications (ed. L. Nachbin), Hermann, Paris, 1975, 277 - 283.

[65] TODD, C., Stone-Weierstrass theorems for the strict topology, Proc.Amer. Math. Soc. 16(1965), 654 - 659.

[66] WARNER, S., The topology of compact convergence on continuous function spaces, Duke Math. J. 25 (1958), 265 - 282.

[67] WELLS, J., Bounded continuous vector-valued functions on a locally compact space, Michigan Math. J. 12 (1965), 119 - 126.

[68] ZAPATA, G.I., Weighted approximation, Mergelyan's theorem and quasi-analytic classes, Arkiv för Mathematik vol. 13 (1965), 255 - 262.

[69] CHERNOFF, P. R., RASALA, R. A., and WATERHOUSE, W.C., The Stone - Weierstrass theorem for valuable fields, Pacific J. Math., 27 (1968), 233 - 240.

[70] DIEUDONNÉ, J., Sur les fonctions continues p-adique, Bull. Sci. Math. 68 (1944), 79 - 95.

[71] INGLETON, A. W., The Hahn - Banach Theorem for nonarchimedean fields,Proc. Combridge Philos.Soc., 48(1952), 41 - 45.

[72] KAPLANSKY, I., The Weierstrass Theorem in fields with valuations, Proc. Amer. Math.Soc. 1(1950),356-357.

[73] MONNA. A. F., Analyse non-archimédienne, Ergebnisse der Mathematik und ihrer Grenzgebiete, Band 56, Spriner-Verlag, Berlin, 1970.

[74] MURPHY, G. J., Commutative nonarchimedean C^*-algebras, to appear.

[75] NACHBIN, L., On strictly minimal topological division rings, Bull, Amer. Math. Soc. 55 (1949), 1128 - 1136.

[76] NARICI, L., BECKENSTEIN, E., and BACHMAN , G., Functional analysis and Valuation Theory, Pure and Applied Mathematics, vol. 5, Marcel Dekker, Inc., New York 1971.

[77] PROLLA, J. B., Nonarchimedean Function Spaces, to appear.

SYMBOL INDEX

INDEX

A

B

C